# The
# Brewsters

## An Interactive Adventure in Ethics
## for the Health Professions

✳ ✳ ✳ ✳

Editors

Jeffrey P. Spike, PhD    Thomas R. Cole, PhD    Richard Buday, FAIA

Story by

Freeman Williams & Mary Ann Pendino

Second Edition, First Printing: April 2012

ISBN 978-0-9854858-2-5

*The Brewsters* story is a work of fiction. Names, places and incidents are either a product of the authors' imagination or are used fictitiously.

Designed and produced for UTHealth and the McGovern Center for Humanities & Ethics by

Archimage, Inc.
4203 Montrose Boulevard Suite 390
Houston, Texas 77006
713-523-3425
www.Archimage.com

"The better part of ethics is simply to consider the effect of your actions on others. The role-playing scenarios in *The Brewsters* provide a great way to do that. Thomas Cole and his associates have devised an ingenious method to turn everyday readers into moral thinkers: have them walk a mile in someone else's shoes – and then what is the reader to do?

"Written to instruct healthcare professionals, this book will be useful to anyone who wants to develop a sharper eye for the moral conundrums of ordinary life and, incidentally, have a good time doing it.

"Here is game playing in the best sense, serious fun."

*Randy Cohen*
*For twelve years the writer of The Ethicist in the New York Times magazine.*

"*The Brewsters* is an edgy, creative, and fun approach to health care ethics. It uses a contemporary format to engage the reader while staying true to form with respect to the rigor of its content.

"The approach represents the cutting edge of how teaching health care professionals is likely to evolve."

*Arthur Caplan, PhD*
*Sidney D. Caplan Professor of Bioethics and*
*Emanuel & Robert Hart Director, Center for*
*Bioethics*
*University of Pennsylvania*


"Brilliant! *The Brewsters* is so much fun to read, you forget you're learning. The innovative choose-your-own-adventure format makes for a truly immersive and memorable experience that will stick with readers long into their careers. All of the scenarios bring up real dilemmas that health professionals will face, possibly in their very first year of training, and this engaging format ensures that students will be equipped to make the right decision when it matters most."

*Kirsten Ostherr, PhD*
*Associate Professor of English, Rice University*

Meet the Brewsters:


Wayne is sure he has osteoporosis

Walter is drunk

Gloria has cancer

Sheila is having an affair.


Choose your own adventure

With three generations of an American family

Getting their health care … from *you*.

# Acknowledgements

This work was created to serve as an introduction to ethics for preprofessional students and students in the health professions. It was initiated as part of The University of Texas Health Science Center at Houston (UTHealth) Quality Enhancement Program for reaccreditation by the Southern Association of Colleges and Schools. *The Brewsters* is a collaboration between faculty at the McGovern Center for Humanities and Ethics at UTHealth and Archimage, Inc. The work was created and edited by Jeffrey Spike, Thomas Cole and Richard Buday. The story was written by Freeman Williams and Mary Ann Pendino. Didactic materials were written by Jeffrey Spike.

We are grateful to many individuals at UTHealth Houston whose contributions made this work possible, including: Nathan Carlin, PhD Assistant Professor in the McGovern Center for Humanities and Ethics; Rebecca Lunstroth, JD, MA, Assistant Director, McGovern Center for Humanities and Ethics; Pei-Hsuan Hsieh, PhD, Education Specialist in Academic Affairs; David Taylor, Director of Education and Technology Services at the Dental Branch, and Pamela Lewis, Administrative Support; Kattie Basnett, copy editor.

# UTHealth Campus-wide Ethics Faculty

Eugene Boisaubin, MD, Medical School

Joan Engebretson, DrPH, RN, AHN-BC, School of Nursing

Cathy Flaitz, DDS, MS, School of Dentistry

Jonathan Ishee, JD, School of Biomedical Informatics

Stephen Linder, PhD, School of Public Health

Dorothy Otto, EdD, RN, School of Nursing

Cathy Rozmus, DSN, RN, School of Nursing

William Seifert, Jr, PhD, Graduate School of Biomedical Sciences

# Preface: Becoming a Health Professional

THE HEALTH PROFESSIONS INCLUDE MANY FIELDS: nursing, medicine, biomedical science, public health, dentistry, and bioinformatics. While each health professional school teaches a different kind of expertise, what runs through all of them is altruism. Altruism means acting unselfishly for the benefit of others. In the health professions, altruism means putting the patient, the public, and the scientific community above your personal self interest. This commitment to altruism involves a fiduciary responsibility to those you will serve — a stringent duty to ensure your decisions and actions benefit their welfare, even if at some cost to yourself. This commitment is codified in the oaths and codes specific to your discipline.

An *oath* (e.g., the Florence Nightingale Pledge for nurses or Hippocratic Oath for physicians) is a formal promise, made in a setting of your peers, to abide by the moral values and behaviors of your discipline. It is a pledge made by an individual, who is then guided by his or her conscience in carrying it out. You may have been required to take an oath or pledge when you entered school, sometimes called an Honor Code.

A *code of ethics* (e.g., Principles of the Practice of Public Health) is a set of rules of conduct that can be enforced by the governing body of a profession or professional school. These rules are specifically expressed as policies and procedures, which are usually binding. One who violates these policies can be disciplined by losing his or her license to practice or being placed on probation, or by being dismissed from school. Once you enter a professional school, you are bound to abide by its oaths and codes and by its policies (regarding, for example, academic integrity, cheating, plagiarism).

Go to the home page of your own school and click on the policies and procedures which govern ethical conduct. Familiarize yourself with them, and remind yourself not only of the consequences of violating these norms, but the highest ethical aspirations they are meant to sustain. The purpose of your school's Office of Student Affairs is not only to enforce standards of conduct. It also exists to inspire, encourage, and embody the ideals of altruism, integrity, and respect.

Act 1 of this course introduces the norms, principles, virtues, and values that are essential to Professionalism. Act 2 is about Clinical Ethics, and Act 3, Research Ethics. In each, you are exposed to ethical issues within and across the health professions. That is, you will be learning about interprofessional ethics.

Oaths and codes are necessary, important, and essential to your moral education, but they are not enough. In school, you are taught professional behavior, compliance, and competencies. Also essential, but still not enough. There is no substitute for cultivating moral virtues and for developing an identity and character that will make you an excellent and successful health professional. It's not only a matter of what you do, it is also a matter of who you are.

Professional identity formation is something that takes place over time. It's a conscious effort to become a virtuous individual committed to your profession. It is not true that you learned everything you need to know about being a good person at home or in church or some religious community. Your beliefs and values are important but they are not universally true, nor should they be the basis of every action. Learning to respect the values and world views of others — and learning how to act when these world views clash — requires both ethical knowledge and a virtuous character.

A *moral virtue* is a habit, acquired disposition, or quality of character that is important for being a good person. You are not born with a virtue. You acquire it, through modeling, reflection, practice, and education. In this course, we will emphasize three virtues which are necessary for being a health professional: compassion, integrity, and respect.

*Compassion* is an emotional and spiritual capacity, a practice of "feeling with" another person. It involves empathy (viewing the world from another person's perspective) and sympathy (feeling pity or concern) for another's suffering or pain. Having compassion for patients and colleagues — even professors — is one mark of a good student.

*Integrity* is also a quality of character, a consistency of action with beliefs, values, and principles. A person with integrity lives in a manner consistent with what she professes and reveals an inner sense of wholeness.

*Respect* is a virtue involving regard for the unconditional value of others. In bioethics respect is built into the duty to acknowledge and defer to the autonomous decisions of patients. Respect also applies to professors, fellow students, and administrators in all professional disciplines.

# Foreword

WELCOME TO *THE BREWSTERS*, a unique course specially designed for students learning about health professionalism and ethics.

You are beginning a journey of professional education that will change you as a person. Your path will require forming a very special identity: someone who works on behalf of others. This means that, along with mastering the scientific and technical material in your field, you will learn to recognize and respond appropriately to ethical problems in healthcare.

*The Brewsters* is a three-part story in which you will become immersed by playing the role of a character. The first part of the story, Act 1, introduces you to the norms, principles, virtues and values essential to Professionalism. Act 2 introduces Clinical Ethics, and Act 3, Research Ethics. Each act exposes ethical issues in your own discipline and in other health professions. As such, you will be learning about interprofessional ethics.

You have a choice of becoming one of two different characters in each act, and you play out the act's storyline from that character's point of view. As you read, you will learn by making choices and experiencing the consequences of your actions. In Act 1, you are a medical student — either John or Cheryl. In later acts, you'll play other men and women: a nursing student, dental student, even a patient and the patient's son.

Act 1 opens with an invitation to a party welcoming new students to the home of clinical faculty, Dr. Enrique Hernandez. Your RSVP sets you off on a journey requiring you to make choices in the story's action. Different choices have different consequences.

In reading *The Brewsters*, you will learn about professionalism and ethics two ways:

1. By immersing yourself as a character in a story, personally experiencing the consequences of your actions and reflecting on those consequences;

2. By reading instructional sections with material you need to master.

Each act is followed by an instructional section providing in-depth explanations of the ethics woven into the storylines. After reading Act 1, read the Professionalism instructional materials. Then read Act 2 followed by the Clinical Ethics materials, and finally, Act 3 and its accompanying Research Ethics materials.

We encourage you to read each act from both characters' points-of-view before reading the instructional sections. See the world through the eyes of health care providers, their patients and their families. Learn from their collective experience. As a health professional student, you are entering an exciting world where appreciating different perspectives is key to understanding ethics.

In the end, we hope this book helps you focus on the ethical and professional dimensions of the health sciences that shape your own story.

# How to Experience This Book

THIS IS NOT A TYPICAL BOOK. Typical books are linear. Reading them is often a passive activity told through a single point of view. This, on the other hand, is an active, participatory story, and it is very much nonlinear. *The Brewsters* is a story told multiple ways through six different characters' points of view.

You might think of interactive stories as a form of video game. Some call branching story path novels, "game books" and credit them as the precedent of today's blockbuster video games. Like any first-person, role-playing video game, *you* play the leading character and *you* decide what your character does in the story.

Instead of chapters, this book is divided into three acts, and each act has multiple scenes. *The Brewsters* is best read starting at Act 1 and ending at Act 3. Although many scenes appear sequential, they should not necessarily be read sequentially. Instead, each scene ends with a choice. After you read a scene, follow a "TURN TO PAGE ..." link to jump to another scene where the action continues. Different links lead to different story paths. You might want to use a bookmark to keep your place in the story.

As you read *The Brewsters*, you will find yourself moving back and forth between scenes depending on the choices you make. Some of your decisions will be good for your character. Others will be bad. It will not matter — whichever way you go, you're bound to learn something interesting. The central idea is learning the meaning of health ethics by experiencing their effects on people, from healthcare professionals, to patients, to families. *The Brewsters* is a "school of hard knocks" story where lessons learned best are the ones characters teach themselves through discovery, trial and error, and chance.

We encourage you to read each of *The Brewsters'* acts multiple times. Select different paths during each read to see what happens. Also choose different characters' points of view. We think you'll find something thought provoking on every page.

# Meet the Characters

*Cheryl Stewart* is a 29 year-old, compassionate medical school student. Born in Atlanta, Georgia, her parents own a modest dry cleaning business and are active in the local chapter of the NAACP. As a child, Cheryl spent many Saturdays with her mother as a volunteer at Victory House, a local free health clinic. She sang gospel in her church's choir and loves music. Cheryl is competent, hard working, and known for her sharp wit. She received her nursing degree five years ago and started working as an emergency room nurse at Austin's Children's Medical Center. Wanting to further her career in medicine, she applied to medical school. To nobody's surprise, she was accepted. Cheryl is now beginning her third-year as a medical student. She is near the top of her class and is widely respected by students and staff.

*John Guerra* is a 25 year-old third year medical student. He was born in Houston to a working class family and received a scholarship to Emory where he graduated with a BS in biology. When he was a child, his sister contracted measles and lobar pneumonia at the same time. Her survival, and the reassuring presence of their family doctor, put John on his chosen career path. John is still occasionally surprised to find himself in medical school. His grades have always been good, but he's had to work hard for them. In his first year, he met Cheryl Stewart, a slightly older student who was willing to argue with him, album by album, who was the better guitarist, B.B. King or T-Bone Walker. After they discovered a mutual admiration for all things Willie Nelson, they became fast friends and have grown closer over the course of their studies.

## ACT 2

*Julie Spates* is a registered nurse working part time in an Austin family medical practice while completing her master's degree in nursing. She is 25 years old, extremely bright and serious. Julie comes from a long line of military medicine. Her father was a dentist in the army. Her mother was a nurse in a M.A.S.H. unit during the Vietnam War. Julie did volunteer work at the Carl R. Darnell Army Medical Center in Fort Hood, sixty miles outside of Austin before she returned to school for her Masters in Nursing. Following her graduation from nursing school, she plans to work in the geriatric clinic at the medical center in Fort Hood.

*Wayne Brewster* is 50 years old. Born in Tulsa, Oklahoma, he is the son of Roger, a machinist, and Gloria, a painter. The family moved to the Austin area in the early 1970s, mostly at Gloria's urging. Wayne grew up suppressing any desire to follow in his mother's artistic footsteps. His father's more utilitarian upbringing left him believing that the job of The Man is to earn a good living for his family. Lacking his father's mechanical abilities, Wayne eventually got a job at the U.S. Postal Service where he remains to this day. He met his wife, Sheila, at a music festival, where certain illegal substances may have been involved. Sheila's pregnancy was unplanned, but the two entered into matrimony with starry-eyed breathlessness — rather against the wishes of Sheila's mother, who thought she was "marrying beneath her." Today, Wayne has a son in medical school, a daughter of high school age, and a young adopted daughter. His father, Roger, died of lung cancer in 1997, but his mother, Gloria, is still living and painting.

# ACT 3

*Parvesh Singh* is a 23 year-old dental student. He and his family moved to New York City from New Delhi when Parvesh was 14 years old, his father accepting a position at the College of Dentistry at New York University. Intensely schooled in the science of teeth from an early age, it was always understood that Parvesh would follow in his father's footsteps — the one concession to youthful rebellion being his desire to open a dental practice as opposed to his father's academic career. Actively hating snow, Parvesh took full advantage of a scholarship to come to Texas for his studies, where the climate suits him much better. Much to his dismay, he is discovering there may be more of his father in him than he thought, as he finds the idea of research ever more alluring.

*Gloria Brewster*, nearly 80, loves to drink, smoke, and curse like a sailor — a Texas cowgirl if ever there was one. Gloria was born in Waco, Texas and earned a Bachelors of Fine Arts degree from Baylor University. She became an accomplished painter before getting married. Although a stay-at-home mom for most of her life, she continued to paint well into her 60s and 70s. Until recently, Gloria had been living in the same middle-class community for the past fifty years. She frequently spends her time playing cards at the neighborhood center after water aerobics in the morning. Although feisty, vibrant and still painting, Gloria is beginning to slow down. When her memory began to fail her, "For everyone's peace of mind," her son, Wayne, asked her to move in with him and his family. Her simple reply: "Heck no."

# ACT 1

"The end aimed at in [ethics] is not knowledge, but action."

*~ Aristotle, Greek philosopher*

*(384 BC - 322 BC)*

# Mail Call

YOU STARE at your university cubbyhole of a mailbox in disbelief. *What is it with these people? Can't they read names?* For the second time in two weeks, someone else's mail is commingled with yours. Well, it's not like it matters. You are all medical school students pretty much getting the same mail anyway. *Hmm, let's see who they think I am this time.*

You find two envelopes, seemingly identical, but with different names. One is addressed to "John Guerra, MS3," the other to "Cheryl Stewart, MS3." You open your mail.

TURN TO PAGE 34 TO READ JOHN GUERRA'S MAIL.
TURN TO PAGE 81 TO READ CHERYL STEWART'S MAIL.

# John: Invitation

To: John Guerra, MS3

Enrique Hernandez, MD

Cordially Invites All Health Science Center

Students

To His Annual

Incoming Class Party and Barbecue

Sunday, August 14

1:00 PM

1314 West 6th Avenue

Austin, Texas

Please RSVP. And, bring your swimsuit!

TURN TO PAGE 35, JOHN, TO RSVP "YES."

# John: Party Animal

SUNLIGHT GLITTERS ACROSS THE POOL like out-of-season tinsel, moving and blending and splitting apart again with the bodies splashing about in the water. You sip your drink and will your body to relax even more into the chaise lounge. The sun will chase you into the cool of the pool soon enough — right now you feel on the verge of a perfect moment, a bit of Hill Country Zen you could carry for comfort into the hectic days ahead.

Then the moment is gone. Cheryl Stewart rises from the pool, still dripping, towel rubbing across her dark skin. She lands in the lounge next to you with an audible sigh. "I don't know about you, but I needed this. Did you need this?"

You smile with what you hope is studied laziness. "I needed this."

She begins to violently buffet her hair with the towel. "You know what I didn't need?" A pause in the hair drying, as she peeks out from the towel. "The first year students keep getting younger."

"It's not that they're getting younger, hon, it's that you're ___"

There is an audible *snap!* as Cheryl brings the wet towel across your chest. "I may be four years older than you, but finish that sentence and I will *finish* you."

"Fair enough," you grin. The wet towel had felt good across your bare chest. The hot Austin summer is doing its best to hang on into the fall, and you know you will soon have to leave this chair, and worse, your drink, to enter the cold embrace of the pool. You look at the pool, crowded with students, some laughing, some splashing, most trying hard to look cool. Some you know, most you don't.

"Have you seen Dr. Hernandez yet?" you ask Cheryl.

"Couple of times. He keeps coming in and out, playing host. Why? You looking forward to his annual big speech to the newbies? Think it'll be the same one as last year?"

You drain your drink. "Oddly enough, yeah, I am. I think he really means all that stuff. At least, I hope he does."

Cheryl smiles in agreement, then the smile suddenly turns serious. "Remember our first year? When Dean Cutter gave the speech?"

"Oh, man. I started thinking about getting a job at my sister's dog grooming business after that train wreck."

"Yeah." Cheryl picks up her drink from the nearby table, swirling it to make sure there is still a bit of ice left. "I'm glad Enrique took that part over."

"Oh, so you're on a first name basis, now?"

Cheryl smiles and sips her drink. "Nah. Someday. I'd like to be." She looks around at the party surrounding her. "I'd like a successful practice to afford a pool like this, too."

You rise from the chaise. "I think we need to get you graduated from med school first."

"Yeah, that would probably help."

You pick up the empty bottle and toss it in a nearby bin. "My turn in the pool. Stand back. There's probably going to be steam when I hit the water."

Cheryl laughs and drains her drink. "Don't give yourself airs."

The water is indeed cold, and you're happy you had the self-control to not yell "Cannonball!" when you jumped in. There had already been several of those, and the genial riot of reciprocation that followed led you to ruefully think that was a game for younger men with thicker skins … and then to even more ruefully realize you were having such thoughts at the ripe old age of twenty-five. *Probably Cheryl's revenge — somehow — for that old age crack*, you think, as you plunge your head into the cooling chlorine.

You surface, trying to do the awesome head toss you'd seen on so many TV shows, and failing. A lap or two around the pool would be nice, but there are too many bodies in the way. You float about a bit, getting splashed and splashing in return, until you feel the need for the sun again … and, to be honest, more drink … and leave the pool.

As you towel off, you walk toward one of the many coolers. The contents are plentiful and varied. Beer, of course. Soft drinks. Sparkling water. You make your decision and reach into the crushed ice, pulling out …

TURN TO PAGE 38 TO PULL OUT A BEER.
TURN TO PAGE 39 TO PULL OUT A SODA.

# John: Beer

… a cold beer …

TURN TO PAGE 40 TO CONTINUE.

# John: Soda

… a cold diet soft drink …

TURN TO PAGE 40 TO CONTINUE.

# John: Photo Op

… and twisting the cap off.

Finally, there, at the house, you see your host: Enrique Hernandez, MD, MPH himself, showing some newcomers out to the pool.

Even dressed casually, Dr. Hernandez always looks professional, you think; the khaki pants and linen shirt wouldn't have looked out of place in some mosquito-netted jungle hospital, and you're fairly certain you have never seen the doctor sweat. As you watch, a short-haired woman only slightly shorter than the doctor comes through the sliding doors and pecks Hernandez on the cheek. Although you've never met her, you recognize the doctor's wife… *Carla, was that it? Yes, Carla.* Dressed in a stylish pants suit, purse over one shoulder, she obviously isn't sticking around. What was it you'd heard about her? *Attorney, yes? Keeps cropping up on the society page.*

Well, why would she want to hang with some mere health professional students? You dismiss the thought with a physical gesture. This is not a day for mean-spirited thoughts and unwarranted assumptions; it is a day to relax, to meet other students, and maybe get a decent buzz on. The time for seriousness is next Monday, when classes resume. Plenty of time for mean-spirited thoughts and unwarranted assumptions then.

As if the universe itself responded, you hear your name. "Guerra! Hey! Guerra, come over here! You too Stewart!"

You finally find the voice's owner, standing waist deep in the shallow end of the pool. *What is his name? Walter, right?* Walter is waving you and Cheryl over. In one hand he has a ludicrously pink plastic camera. Walter is motioning to a group forming a few feet away in the pool. "I need a picture! Come on!" Several others start motioning and calling. It is, frankly, the women in their bikinis that prompt you to climb back into the water and join them.

Walter motions everyone to get closer together. "Everybody got a beer? Good."

One of the girls who had helped you make your decision looks at the cup in her hand. "Rum and Coke okay?"

Walter grins. "Even better." He begins lining up the camera as you approach the group. "Okay, everybody, this is for my Flickr page. I need to show all my old frat brothers who are getting their MBAs that they made a poor life choice, so I need everyone to look like we are at the Party of the Century. It's my hope to make them want to slash their wrists with their spreadsheets." Walter looks up from the camera, a twinkle in his eyes. "So if any of you ladies want to show a little extra flesh, please feel free." He returns his attention to the camera. "Hurry up, John and Cheryl. I want to see drunk and lecherous on the count of three! One! ... Two! ..."

TURN TO PAGE 42 TO JOIN THE GROUP PHOTOGRAPH.

TURN TO PAGE 48 TO DECLINE THE PHOTOGRAPHY SESSION.

# John: Cheese

YOU PRESS YOURSELF into the mass of flesh and look at the camera on "3!" There's a flash, surprisingly bright in the midday sun. "Okay," says Walter, "you guys didn't look drunk enough. Try again, and get closer together."

There's some generalized splashing and you feel more bodies pressing around you as Walter begins to count down again. "Come on! That's got to be more enjoyable than you're making it look! One! … Two! … Three!"

The crush around you begins to relax, but Walter's still making like some clichéd tyrant film director. "More wanton! Think *Health Students Gone Wild!*" and the crush returns around you. There are even a couple of calls of "Woo hoo!" Blue blobs are dancing before your eyes from the flash, but you're more than aware that some very attractive young women are pressing themselves against you. One throws her arms around your neck and plants an impulsive kiss on your cheek. You can't help grinning like a madman. A somewhat embarrassed madman, but that appears to be what Walter wants as the flash goes off again.

The tangle of bodies loosens itself. You cast about for the girl who grabbed you, but whoever she was, she's gone, off with someone else or in one of the giggling clutches trying to peer over Walter's shoulder at the camera. You blink rapidly, trying to get rid of the flash spots obscuring your vision. One more photo, you think, and you might not have been able to find your way out of the pool.

*Well, I'm glad that's over,* you think. *I don't think I've done anything that juvenile since my freshman year ...*

TURN TO PAGE 44 TO CONTINUE.

# John: Flickr

TWO WEEKS LATER, you predictably have your hands full. You're certain there are people somewhere in the universe with busier schedules, but you can't find it in your heart to feel sorry for them.

There's something else, too. You're seeing a lot of knowing smiles and hearing an occasional "Hey, Tiger!" or similar remark — like there's a secret organization of people around you that know something you don't. But, you also believe it's too early in the year for stress-related paranoia, so you mentally shrug it off and try to bury yourself in your studies.

Until you get a comment muttered to you in passing while you're talking to Cheryl in the hall. "Jerk."

Looking at the nursing student as she walks away, you turn to Cheryl. "Do you have any idea what that's about? I've been getting it a lot, lately."

She looks at you pityingly. "You mean … you don't know?"

"Know what?"

She pulls you into an empty room and begins setting up her laptop on a desk. As it boots, she asks, "So you didn't get the email from Walter?"

"Walter? I generally just delete those. If he's not trying to set up a party, it's all whining about first-year stuff that doesn't concern me."

"Oh, this might," she says, angling the screen toward you. You instantly recognize the thumbnail photo layout of Flickr.

You had an account back in the undergraduate days. You haven't paid much attention to it since …

Wait a minute. You recognize some of those people in the picture. It takes you a second longer to recognize yourself, you're grinning so widely in the photo, showing none of the embarrassment you remember. Well, now you know what the girl who kissed you looked like. You don't remember putting your arm around the girl in front of you, your beer bottle dark against her belly. And you certainly don't remember her grabbing her top and …

"Oh, wow …" you say.

Cheryl considers your witty reply. "Yup, *wow* is a very good reaction."

"I … I had no idea that was going on."

"That's another possible reaction. Frankly, I liked *Oh, wow* better."

"I've got to get him to take that down!"

"I don't think you're the first one with that reaction, John, but, I'm afraid the damage is already done. Let me show you what a Google search for your name brings up …"

TURN TO PAGE 46 TO CONTINUE.

# John: The Letter

BACK IN UNDERGRAD, you recall there was one English professor — Thomas, the guy who always impressed you as looking like a shop teacher who had mistakenly put on a suit that morning — who kept harping on the importance of symmetry in great literature. That the closing event of a story should mirror the starting event.

Well, there's your mirror event, sitting in your mail cubby. An envelope, looking very innocent except for the return address, it's top line in bold, impressive copperplate print: **OFFICE OF THE DEAN OF STUDENTS.**

Staring at it isn't going to improve it, you decide, and you pull it out and rip it open swiftly, hoping for a quickly pulling-off-the-band-aid effect.

> Dear Mr. Guerra:
>
> It has been brought to our attention by the parent of a student that a certain series of photographs taken during a party at the beginning of the semester have been published on the Internet. These photographs document activities and behavior that do not reflect the image and reputation of our institution and its student body, of which we are justly proud.
>
> Therefore, this Office has no choice but to launch an investigation into these

photographs and the people responsible.
Please make an appointment with the
Dean of Students as soon as possible.


There's more, but you're concentrating on the plummeting feeling in your gut. This is not going to be pretty.


TURN TO PAGE 40 TO TRY AGAIN.

# John: Camera Shy

YOU LAUGH AND QUIETLY SWIM away from the group, hoping that the laugh will soothe any hurt feelings. Lifting yourself up to sit on the edge of the pool, you dry off and head toward the hors d'oeuvres table in the house for some cheese and crackers, instead of the cheesecake you just passed up.

A moment later, you spot Cheryl. "You didn't want in the group shot either?" you ask.

"Group shots I'll do. What Walter wants is an orgy."

You glance back at the group in the pool. Walter is calling for a second shot. "More wanton! Think *Health Students Gone Wild!*"

"Why didn't *you* do it? Not enough girls in there?" Cheryl is smirking.

You laugh. "Enough for me. Not for Walter, apparently." You shrug. "Didn't feel right. Maybe it's something I would have done back in my freshman year, but ..."

Cheryl laughs. "Let me tell you a secret. Not *even* in my freshman year. And when I am in one of these photos, I try to make sure I don't have a drink in my hand, or anything like that."

"Really? Why?"

"Something I noticed back in *my* freshman year, genius. My roommate was laughing and showing me this website with all these photos of drunken girls, and I realized that they had all been taken off personal sites and whatnot."

"Yeah, I think I've heard about things like that. Something."

"Well, once stuff like that is out on the Internet, it stays out there forever. So if something like that ever shows up — particularly if it shows up after I've got that 'Doctor' in front of my name? No way, Jose. I want to be known as 'The Sober Girl On The End.'"

"Smart." You put on your best serious face. "Is this the wisdom that comes with age?"

She gives you a look you have seen a thousand times. "You know, I bet there is enough going on in that pool that I could drown you and get away with it."

You are saved from certain death by an older woman in a wildly colored sundress, a wide-brimmed hat protecting her fair skin from the Texas sun. Her ringing call of "Hi-iiiiiii!" precedes her like an opening salvo from an invading force. She walks toward the two of you, her hand outstretched, a thousand-watt smile on her face. In all of this, what you notice is the drink in her unextended hand, a bright red concoction with a green paper umbrella. *Where did she get that umbrella?* is the insane thought that crosses your mind, and then she is upon you.

"I'm Leslie Crenshaw, Dr. Crenshaw's wife? I'm so happy Dr. Hernandez asked us out here this year! We never seem to go out, it's all work, work, work, and study. You know how it is, right?" Both you and Cheryl inhale to answer, but she goes on, "This is a great party! Isn't this a great party? I really miss socializing like this. I mean, my book club is great, and they're all great people, but every now and then you've just got to go out and expand your horizons, am I right? I mean, you just stagnate, am I right? I'm really enjoying meeting all the other doctors out here!" She stops abruptly, obviously waiting for one of you to reply. At this point, both you and Cheryl are almost afraid to — but as you're about to say something, she continues, "So! What kind of doctors are you?"

Leslie smiles, waiting. You find yourself sharing her smile, but for a different reason. *Well, this could be fun*, you think. *What kind of doctor should I be? Laser oncology? Quantum neurology? Veterinary osteopathology?* You open your mouth to reply …

TURN TO PAGE 51 TO INTRODUCE YOURSELF AS, "DR. GUERRA."

TURN TO PAGE 55 TO INTRODUCE YOURSELF AS, "JOHN."

# John: Dr. Guerra

"OH," YOU SAY NONCHALANTLY, "I'm a doctor specializing in reproductive endocrinology."

"Oh," says Leslie, "then I really need to speak with you, maybe get a card or something?" You open your mouth, but she charges on. "You know, Lawrence and I have been trying to have a child for ages and so far we just haven't been able to. I mean, there's been a couple of times … but at this point I've had two miscarriages, and really, one was devastating enough, for both of us." It's almost frightening how subdued she's become, compared to how this conversation began. "I mean, it's never just a baby, you know? You're planning the first birthday party, her first prom, where she's going to college. Lawrence is afraid to even try again, so I've been reading up on this endocrinology stuff when I can, and I was thinking …"

You finally hold up your hands and say, "Ma'am! Ma'am, I'm sorry. I'm …" and you realize with a sinking feeling there is no easy way out of this. "I'm not actually an endocrinologist. I'm just a student. I was …" and you realize the word does not excuse it: "… joking."

"Oh," she says, and somehow that has become the worst word in the world; that breezy, chatty woman who practically bowled you over with her force of personality seems to cave in on herself, the light shining from her eyes dulled, except for a slight glimmer of tears.

You say, "I'm sorry," again, even knowing the repetition doesn't help.

"That's … okay," she finally says. "You couldn't have known." And with that said, she turns and walks away, her gaze distant, her mind obviously somewhere else.

Okay, you think, that was stupid. I'm glad it's over.

Then you hear a throat being cleared behind you. You turn to see Carla, Dr. Hernandez' attorney spouse, standing nearby, glaring at you.

Oh, crud. *I thought she was gone.*

TURN TO PAGE 53 TO CONTINUE.

# John: Overheard

THE TWO OF YOU STAND FOR A MOMENT, looking at each other. Carla finally breaks the silence. "If this was an after-school special, this would be the point where one of us would say, 'Well, we learned something here today.'" If possible, her gaze becomes more steely. "Did we learn something here today?"

"Well, I …"

"Allow me." She steps closer to you. "You learned you're working toward a profession where you can assume nothing. You learned that something you think is funny could honestly, truly hurt someone in a way you could never suspect. Above all, you'd better have learned that calling yourself a doctor before you have that all-important degree is stupid, wrong and actionable."

The ways she says *actionable* puts you in mind of gavels banging, cell doors slamming shut, guillotines whooshing toward necks.

"Yes, ma'am," you reply, knowing it sounds as weak as you feel.

"She was on the verge of telling you her health history, and no amount of 'sorry' could have gotten you out of the world of hurt that would have caused. You know who I am, right?"

"Yes, ma'am."

"Good. Then you know I'm an attorney. And you do not want someone like *me* getting a hold of someone like *you* over something like *this*."

"No."

She looks at you a moment. "Leslie Crenshaw's a friend of mine. My first impulse is to gut you like a catfish, but you did seem genuinely sorry." You make a mental note to check your skull for holes later, because you swear her eyes are shooting laser beams through you. "We are never going to need to have this conversation again, are we?"

"No. No, we're not."

"Good. Have a nice day."

Carla turns and walks into the crowd looking for Leslie, leaving you feeling distinctly small, and, at best, four years old.

TURN TO PAGE 48 TO TRY AGAIN.

# John: Not Yet

"OH, NOT 'DOCTOR' YET," you say. "That comes after my medical degree, not before. We've both got two years to go, although it feels like a decade."

Leslie's smile, if anything, becomes more enthusiastic. "Why, isn't that great! It must be an exciting time for you both! I can only imagine! Well, hang in there!" She's already spied another group a bit farther off, and she glides toward them, free hand waving, another "Hi-iiiii!" lancing through the air.

You and Cheryl are both silent for a moment, looking at her move away.

"Wow," you finally say.

"Yeah," Cheryl responds. "That woman is a force of nature."

You hear a voice behind you say, "Excuse me!" You both turn to see a young man with dark, wavy hair, and skin even darker than Cheryl's approaching. You recognize him as one of the students in Walter's group shot, but that's about it. "May I ask you two a question? Where do you guys know Mrs. Crenshaw from?"

"Uh, we ... don't," you reply. He sees your quizzical looks.

"Oh, sorry. Hi, my name's Parvesh. In the dental program? The Crenshaw Center?"

Many things are suddenly clear. "That's right! Dr. Crenshaw's dental! I thought I saw him around."

Parvesh nods. "He told some of the dental students about the party, said there would be a practicing dentist or two here, if we had any questions, you know?"

Cheryl nods. "Dr. Hernandez works with many local doctors. I've seen some show up. I'm Cheryl, incidentally. And, although John is not yet a doctor, he *is* a Neanderthal and, thus, incapable of introducing himself."

"Pleased to meet you both," says Parvesh.

You are too busy trying to rouse your memories of last year to respond to the gibe. "Wait, there was this time I met a dentist who did some consulting work with Dr. Hernandez... Nguyen, I think. She was what? Japanese?"

Parvesh smiles. "With a name like Nguyen, I don't think so. Vietnamese, surely."

Cheryl's tinkling laugh cuts your reverie short. "Like I said: Neanderthal."

You take a sip from your drink. "Anyway, I haven't seen her here today."

"Well, I wasn't going to ask about her, anyway," says Parvesh. "You two look like you've been around awhile." You both cautiously nod, with Cheryl tossing you one of *those* looks. "I was wondering ... um, who is that?"

You and Cheryl turn your heads in a synchronized move following Parvesh's pointing finger. Dr. Hernandez is talking to two women near the house. One is a statuesque redhead, wearing a large blue cambric work shirt open over her modest black bathing suit. Her oversized sunglasses complete the illusion that Hernandez is chatting with a movie star. The other is dressed like all the other students at the party. Leslie Crenshaw soon joins the crowd.

"You mean Julie?" you say. "Julie Spates? She's a nursing student. Actually, like Cheryl here, she's an RN. She's studying to be a nurse practitioner now."

"I'll bet that's not the one Parvesh is referring to, John," Cheryl notes, turning to Parvesh.

"The one with the hat." Parvesh has a wide smile on his face.

"Oh, that's Dr. Frey," Cheryl says. "She's kinda new. Just took over as the university's new RIO. I hear she's a MPH with an informatics degree."

"Yeah," you turn too, not wanting to stare. "Margeaux Frey, right?"

"Right. Your memory is improving."

Parvesh looks puzzled. "MPH, I understand. Masters of Public Health. What's a RIO?"

"Research Integrity Officer," Cheryl says. "Believe me, you don't want to meet her under the wrong circumstances."

Parvesh smiles and then looks furtive, as if telling a secret. "Maybe not, but have you heard her French accent?"

"French Canadian, I think." Cheryl looks at you. "Hence, the Margeaux."

"It's enough to make a man melt on the spot."

Cheryl clears her throat. "Okay boys. Never mind her. Who's *that?*"

An older gentleman has joined Hernandez and Frey, but can only be judged as "older" due to the sprinkling of gray in an ample mane of black hair. He obviously works out and isn't afraid to show it, wearing only his swim trunks and sandals. He has a bottle of wine in each hand and is apparently asking the other two doctors' opinions.

"Oh, that's Dr. Cochran," you tell her. "Clinical faculty. Surgery, I'm almost sure. I think he even runs the surgical residency program."

"Hmm," says Cheryl. "I'm thinking about a surgical residency."

Another gentleman walks into the group. Older and more distinguished in both dress and demeanor, and puts his arm around Leslie Crenshaw. You recognize Dr. Crenshaw immediately. He and Cochran are obviously friends, each slapping the other on the shoulder in gentle laughter.

A decision was apparently rendered in the doctors' wine conference, as Cochran walks into the house, perhaps looking for a corkscrew or something. Seeing Cheryl watch him intently, you lean toward her. "I don't think he has a French accent, Cheryl."

"I couldn't care less about his accent, John." She turns to you and Parvesh. "I'm going to leave you two boys to your fantasies about French women. I'm going to get a refill."

As Cheryl walks away, Parvesh moves closer. "So what do you think? You think Hernandez and Frey have something going on?"

"What? Dr. Hernandez and …" you look back at the two. Hernandez has obviously just said something humorous, as Frey laughs, a great uninhibited gush of bell-like peals. You consider it a second, then turn back to Parvesh.

"I doubt it," you say. "He's always seemed like a straight arrow. Probably has enough on his table with his practice and all … not to mention he's married to an attorney. I would not like to deal with that."

"Soooo … what you are saying is, there is a chance for a first year dental student, right?"

You look at him for a moment, wondering if he's kidding. He suddenly grins, and you realize he is. It's your turn to laugh. "Dream big, my friend. Dream big."

Parvesh joins in the laughter. "Ah, well. Maybe I'll just suddenly go for an informatics degree, too. I would need advice, after all."

You notice a petite Asian woman coming through the house's sliding doors, waving to Hernandez and Frey. You point her out. "That's your Dr. Nguyen, the practicing dentist. You might want to pay more attention to her than Dr. Frey right now."

Parvesh smiles again, easily. "Right. I'll get Crenshaw to introduce me. Thank you, John."

"Anytime. See ya." You saunter to a cooler, waving to the occasional familiar face. There is one at the cooler. Walter, a half-empty beer in one hand, that stupid pink camera in the other. He's paging through the pictures in the preview for the benefit of a couple of admittedly cute girls on either shoulder. He glances up from the camera once the show is over. "Hey, Guerra!"

"Hey Walter. You get the shot you wanted?"

You open the cooler and Walter joins you. "Aaaaah, it'll do. Haven't got the killer one yet, but …" he drains the beer, lobs it toward a bin, then almost instantly bends down to retrieve another one. "I'm sure as the day wears on, and the demon drink does its job ..."

"Uh huh," you grunt.

TURN TO PAGE 61 TO CONTINUE.

# John: Refill

YOU LOOK INTO THE COOLER. There's more beer of course, Walter hasn't begun to put a dent in that supply yet. Soda. Some water.

That makes you look up and see Walter chugging down the contents of that last bottle, as if trying to prove a point to any audience that might be watching. Well, he's not feeling any pain now, you think. It's been years since your last hangover, and you see no reason to break that particular streak.

You pull out some vitamin water and wonder for a moment if what you're feeling is triumph over being so adult, or regret over being such a stuffed shirt.

Whatever. *I'm the guy who won't be hugging the toilet later.*

TURN TO PAGE 62 TO CONTINUE.

# John: Is it Cheating?

YOU HEAD BACK TOWARD YOUR LOUNGE CHAIR, catching snippets of conversations along the way. One discussion in particular draws your attention: Kyle, a second year student, is holding court to a group of first-year students. It looks like a group of Boy Scouts listening to a ghost story around a campfire. You soon realize that's not far from the truth.

"And that guy Linderman?" he is saying. "Take every by-the-book, no-talking-during-class, read-one-hundred-pages-on-the-first-day-of-class teachers you have ever had, multiply it by ten, and you've got Linderman."

There are a couple of audible moans from the group. Kyle seems to be enjoying this. "Don't take my word for it, Cheryl had him. Am I right, Cheryl?"

You didn't even notice her wander over. Cheryl shrugs. "He's tough, but fair."

"Yeah, if by fair you mean a fair amount of sadism thrown in," says Kyle, reasserting control of his audience. "His tests are the stuff of legend, and the final ..." he fakes a shudder.

You settle in. This is getting entertaining. You're beginning to wonder if he'll end the story with a sudden scare just to see if anybody jumps.

Kyle continues. "But here's the thing. That big, bad, final exam? It's the same every year. Not sure if he doesn't want the extra work to make a new one, if he knows he can't make it any harder, or what ... but it's the same test year after year." He leans forward. "When I was in your place, I was kinda freaking out. Felt like anything less than total perfection would kill my entire career, you know? Linderman has a way of making you feel like that. So this second year student told me about the final exam, and said, 'Hey, relax, I memorized all the questions and wrote them down after the test. Take a look at it.' And I gotta say, it was a big help to me."

Kyle leans back. "So if you start freaking out, don't worry about it. I still have that copy, so I can carry on that grand tradition."

TURN TO PAGE 68 TO SAY SOMETHING ABOUT THIS.
TURN TO PAGE 64 TO JUST LET IT GO.

# John: Let it Go

"AS A MATTER OF FACT," SAYS KYLE, rising, I have a little something for you to start out your year. Party favors."

"Party favors?" asks one of the students.

"Over the summer, I had to make copies. I thought to myself, I'm already slaving over a photocopier. Why not run off some copies of Linderman's test questions too? So, as fate would have it, I have a bunch of them sitting in my car. Anybody who wants one, follow me."

Kyle starts toward the house, a number of the first-years in tow.

TURN TO PAGE 65 IF THIS IS SOMETHING YOU'VE GOT TO SEE.

TURN TO PAGE 70 IF YOU FEEL SOMETHING IS WRONG WITH THIS.

# John: Kyle Distributes Test

*THIS,* YOU SAY TO YOURSELF, *I GOTTA SEE.* Kyle exits the patio gate near the house. You follow the small mob to Kyle's car. He opens the trunk and reaches in. You half expect him to shout, "SUCKERS!" and turn to drench his followers with a Super-Soaker or something. But, no, he pulls out a cardboard box filled with papers.

"I thought of having them all rolled up with blue ribbons like diplomas," he says, "but I ran out of time."

Kyle starts distributing stapled sets like pamphlets in the commons. The same curiosity motivates you to step forward and take a copy — and yes, it does look like Linderman's test, at least as well as you can recall.

"Look familiar?" Kyle smiles.

You are about to answer when you hear a voice above the murmur of Kyle's followers.

"What's going on here?"

The small crowd parts to reveal Dr. Hernandez.

*Oops.*

TURN TO PAGE 66 TO CONTINUE.

# John: Oops

HERNANDEZ TAKES ONE OF THE SETS from a nearby student and looks at it. Then he looks at Kyle, who suddenly seems as if he wishes he were not holding that particular cardboard box.

Dr. Hernandez scans the first page, then flips to another … and another. "Is this what I think it is?"

There is some uncomfortable shuffling and silence.

"That's what I thought," says Hernandez walking toward Kyle. "Hand me the box please."

Kyle complies and Hernandez turns to the others. "Everybody, put those copies back into the box."

Silently, their faces muffled by shame, everyone, including you, places the photocopies of the test questions into the box proffered by Hernandez.

The doctor begins to walk back to his house. "You know, when I see students clustered around a car, I feel the need to check it out. Make sure it's not drugs, or anything that can get them in trouble."

He turns back to you. "Frankly, this is just as bad. This is no way to start a journey that should end with each of you fully becoming a professional." He nods toward the contents of the box. "This is the opposite of professionalism. This is taking the lazy, unfair, unearned way. I'm afraid each of you is going to have to talk with the Dean of Students about this."

That stirs the group as Hernandez turns and continues toward the house. Several cast resentful glances at Kyle, a few head after Hernandez.

You? You don't feel much like partying anymore.

TURN TO PAGE 62 TO TRY AGAIN.

# John: It's Cheating

YOU SAY, "COME ON, KYLE. It wasn't that hard. I did okay on it without a cheat sheet."

Kyle responds, "I'm not saying they should, like, copy it down on their shirtsleeves or something. I'm just saying it might be helpful for them to know what's on the test."

"If that's not cheating, what is it?"

Kyle looks exasperated. "Look, whatever else Linderman is, he's a smart guy. How could he not know students are doing this, if he just copies and pastes every year?"

"It's operating with an unfair advantage. It's cheating."

Kyle shakes his head. "I think we're talking about a very gray area."

Jessica, another second year student, speaks up, "You want a gray area? I got one. I want to know what you guys think about this — I just read about it a few days ago."

She's a natural storyteller. The original newspaper account can't be anywhere as good as this.

"So," she continues, "these guys are taking microbiology and the professor wants them to act as a team, right? Learning together, working together, doing everything but living together. And, one of the girls did the full secretarial thing in high school, right? Typing and I guess, shorthand. Because her notes are incredibly detailed. So she emails everyone in her class, sends them her notes. Everybody's on the same page."

"Literally," says a member of the audience.

"Right, but, here's the kicker: the Dean's office gets wind of it and they declare that the entire class is cheating."

You have to admit, the wave of consternation and shock that washes through her congregation is delicious. Their jaws drop.

"What? They were just doing what the professor said!"

"Maybe they were doing what he said a little too much."

"Well, if the other students were, like, not taking notes at all, slacking off, then I could see it."

Jessica shrugs. "As we've seen with these two," she says, indicating you and Kyle, "what's reasonable can be up for grabs, depending on what side you're on. Is the whole cheating thing subjective, then?"

Kyle's back on stage. "And speaking of up for grabs ... I have a little something for you to start out your year. Party favors."

"Party favors?" asks one of the students.

"Over the summer I had to make copies. I thought to myself, I'm already here, slaving over a photocopier. Why not run off some copies of Linderman's questions too? So, as fate would have it, I have a bunch of them sitting in my car. Anybody who wants one, follow me."

Kyle starts toward the house, a number of the first-years in tow.

TURN TO PAGE 70 TO CONTINUE.

# John: Report Kyle

YOU WATCH KYLE HEAD TOWARD HIS CAR, a number of the fresh students in tow. This eats at you in a way you're afraid is selfish. After all, nobody offered you a copy of that test your first year. Is what you're considering because of that?

These are big questions, you decide. What you need to do is talk to somebody about it — sometimes just the act of doing that makes things clear to you. Cheryl's nowhere to be found. Most of the people around you are strangers at this point … and with that moment of clarity you were praying would come, you realize who you need.

It's a little calmer inside the house, and it only takes you a few moments to find your goal: your host, Dr. Hernandez, in conversation with a couple of other faculty members you barely recognize. You insinuate yourself into their group a bit tentatively. "Excuse me, Dr. Hernandez? Can I speak with you for a moment?"

Unruffled, the doctor turns to you. "Why sure," he says and excuses himself to the other two. He walks with you a few steps away so you have a bit of privacy.

Quickly, you outline to him the situation with Kyle, his distribution of the tests, and your confusion over what to do. Hernandez' face darkens a bit at the news, but his face softens as he considers you.

"I'll go outside and check it out," he tells you. "You've done the right thing, John. This is definitely cheating, and it's something that needs to be dealt with quickly. Now what I think you should do, come Monday morning, is report this to the Dean of Students. Tell him exactly what you've told me. I'll probably be able to back it up."

"I should do that? I don't want to get anybody in trouble …"

"You're not," Hernandez says quickly. "Kyle is. He's an adult, John, and he wants to be a doctor someday. If he makes this choice now, this bad choice, what does that say about his character? About any future choices he might make? Maybe it's a mistake, maybe he can straighten out, or maybe he just needs a little extra guidance. Or, maybe a little scare. Or, a good talking to. But we can't — the Dean can't — correct his path unless somebody brings it to his attention before it's too late. That someone is you, John. It's not easy, I admit, but it's right. Nobody ever promised you that becoming a healthcare professional was going to be easy."

"Yes, sir. I see what you mean."

"Good. I'm going to check out front. Thanks again, John. You're doing a good thing."

TURN TO PAGE 72 TO CONTINUE.

# John: The Speech

A BIT LATER, your attention is caught by a sudden racket. Casting about, you find its source: a fourth year student near the house, knocking two empty wine bottles together; careful not to break them, but producing a clear, remarkably loud ringing sound. Nearby, Dr. Hernandez is climbing a stepladder brought from the house to give him some extra height. The music and general hubbub quiets as head turns toward the doctor, who stands with his arms outstretched, motioning for quiet.

"If I could have your attention for just a minute, please," he calls out, his voice strong and loud without shouting. "I have a few things I'd like to say."

The party quiets even more, a few people stroll toward the house to get a better view. The bathers empty the pool, towel off, and head into the crowd. "First of all," he continues, "I'd like to say we are very pleased with this wonderful new class, chosen so carefully by our Admissions Office."

There is a boisterous, "YEAH!" from the back of the crowd. A quick glance proves the source to be Walter, with (*of course*, you think) a beer in his hand.

"And so enthusiastic," smiles Hernandez. A general chuckle runs through the crowd.

Hernandez goes on, "Your application essays were read, your interviews were considered, and it was felt that each of you had the makings of a good healthcare professional. You were all tops in your class, but what is more important, you all had the sort of character we want to be associated with our school."

You glance quickly back at Walter, but he, like the rest, seems to be listening raptly.

"Your essays told us of a deep motivation to relieve suffering, to research and cure disease, to be a positive force in the world. Reading those thoughts makes me proud of my profession, and proud of you, who will become my colleagues."

Hernandez changes his stance slightly. "You've got something of an adventure ahead of you. When you applied to this school, you weren't just trying to get into a highly competitive graduate educational program. No, it was like you were proposing marriage to someone."

There is a general wave of laughter through the crowd. Hernandez holds up a hand. "No, no, hear me out. When a school interviews you, it's like a first date. They're finding out what kind of person you are, if they want to be seen with you in various settings. When we admitted you, we accepted your proposal. We didn't just like your grades, we liked *you*.

"Now, once you were admitted, you still had the chance to change your mind. Some people get cold feet when they're faced with the cost and complexity of a wedding. But, if you don't back out, if you continue, you are making a lifetime commitment to another person. You care what they think about you, and you're willing to make some personal changes to meet their expectations. Becoming a healthcare professional is not just getting a degree, or learning a lot of new information. It also means a change in your identity. First, others will start to see you differently. Later, you will start to think differently, and finally, you're going to see yourself differently as well."

He gestures to the crowd at large. "Most of you here are entering medical students. I want to point out that we've also invited some of our more advanced students, up-and-coming second and third year folks, so you can get to know them, and maybe even be gifted with some wisdom about the weeks and months to come."

He makes a show of looking over the faces before him, pointing to various people. "I also want to acknowledge that we've got some colleagues with us, students and faculty from the Dental School, Nursing School, the Graduate School and … there you are! The School of Public Health! I'm glad all of you could make it!"

You survey the unfamiliar faces in the crowd.

"So, you may be asking yourself, 'Why the eclectic mix?'" Dr. Hernandez smiles in a dramatic pause. "Because I wanted to underscore a fundamental truth about our health professions. All of us — physicians, nurses, dentists, researchers, public health and bioinformatics professionals — are part of the same team."

Hernandez squints as he surveys the crowd. "I also see a few of my patients here today too, but don't worry, I won't identify you. HIPAA, you know." A smattering of laughter follows.

Hernandez glances down at his feet, gathering his thoughts, then looks back up. "If there is anything you take from me standing up here and shouting out at you, it's this: It is important to remember that we, all of us, every school, every discipline, share a common goal: to improve the quality of healthcare, and its delivery. And, it is equally important to realize that none of us can do that alone. We can only be effective as a team. So please, say 'hi' to someone you don't know today. Get to know your team. Again, welcome."

Hernandez steps down from the ladder to no small applause. You find yourself nodding appreciatively. That actually *was* better than last year's. You decide it is a good idea to mingle and meet some of the other schools' students. Parvesh seemed okay, and it looks like some of his classmates are setting up a volleyball net at one end of the yard.

Yeah, you're definitely glad you came.

$$* \quad * \quad * \quad *$$

After several rounds of volleyball you're feeling pretty good. You can still spike with the best of them. Dr. Cochran was a surprise, though — that guy plays for blood. You're going to feel this in the morning, but it's a good burning sensation in muscles too long unused.

The sun sets during the game. The party is slowly winding down, the music and conversation are more subdued. Probably time to pack it in, you think. Tomorrow's going to be rough enough as it is. As you catch your breath, you wish for a breeze to help evaporate the sweat from your body. The weather's not cooperating though, and your eyes fall on the pool again. Less crowded, more sedate. One last dip in that cool blue water, and you'll be ready for the drive home.

Oh, yeah, the water's definitely still cold enough to take your breath away. You splash across your face, still feeling a little pull and weakness in your arms from the game. You start looking for the good doctor's hot tub — you're going to have some sore muscles in the morning. Some of that hot water jetting over your shoulders will help you like yourself better in the morning.

You paddle to the end of the pool closest to the hot tub and note that it's occupied, but not terribly crowded. The relaxing hot waters have done their job, though, the people in there are much more sedate and quiet … and then you notice Cheryl.

You'd wondered where she'd gotten to after Hernandez' speech, and now it appears you know — she's in the hot tub, and she's very close to Dr. Cochran, that surgeon she was eyeing earlier. *Too close*, you think.

As you watch, Cochran whispers something in Cheryl's ear, one arm reaching around her shoulders. Cheryl laughs easily, unconcerned by the contact. Then Cochran, his face still close to her ear, kisses her on the cheek.

You are starting to get uncomfortable with this. You look away, trying to organize your thoughts. Should you say something? Climb into the hot tub with them, maybe try to break it up? On the other hand, Cheryl's a big girl. She's not exactly screaming for help. She might resent you trying to interfere in her life like that, and you like her too much to risk that. Then again … Why does everything have to be so complicated?

Some would call it discretion, some would call it cowardice. You climb out of the pool and dry off. It's time to head for home, where things are simpler.

TURN TO PAGE 77 TO CONTINUE.

# John: Drunk Walter

DRY AND DRESSED, you say so long to the new students you've met throughout the day — the ones still hanging around, at least — and thank your host. You're a bit surprised to find yourself heading for the front door at the same time as Cheryl, but you also think it's a good idea to simply not say anything — which should stand as a milestone in your relationship with her.

As you both walk out the front door, the first thing you notice is Walter standing in the street next to what is apparently his car. He is pawing through a duffle bag and an open bottle of beer sits on the car's roof.

"Oh," says Cheryl, "this is going to get very sad, very quickly."

You both approach him as his search through the contents of the duffle bag becomes more frustrated. "Hey, Walter," you call out to him. "¿Que pasa?"

He pauses his rummaging to look up at you. "Oh, hey." He starts patting on the pockets on the side of the bag. "I'm jus' lookin' for the keys to my chariot, y'know? I mean, I gotta have 'em. I got here, right?" He suddenly puts his hand in his pants pockets and laughs, a sputtering giggle. He pulls the keys from his pocket. "Las' place you always look, am I right?" He attempts to sort through the keys and instead drops them on the lawn. "Aw, crud."

As you watch Walter laboriously bend over to retrieve his keys, you realize you're not quite able to retreat to the simplicity of your apartment yet. Walter's been binge-drinking all day long. There's nothing wrong with having a drink every now and then to relax, but Walter seems determined to burn out his liver before school even starts. Walter might bear a bit of watching in the coming year — his current state might be a warning of future problems, maybe even a continuing problem the Dean should know about.

Of course, you've got the problem right now. "Walter," you say, "I'm not sure you should be driving."

"Nonsense!" Walter shouts. "Drivin's like autopilot to me. I'll be fine ..." and in the course of gesturing, his keys go flying into the lawn. "Aw, crud," he repeats.

TURN TO PAGE 79 TO TAKE WALTER'S KEYS.
TURN TO PAGE 128 TO LEAVE WALTER TO HIS OWN DEVICES.

# John: Walter's Keys

YOU HAVE A SLIGHT ADVANTAGE OVER WALTER, and snatch up his keys from the grass. Walter holds his hand out for them. "I'm serious, Walter. You can't drive."

"Yes, I can," he slurs, trying to sound sober. "Just gimme my keys."

"No. Either I'll take you home, or I'll call you a cab. That's your choice."

Walter advances on you, and he's starting to look menacing. "Will you jus' gimme my dang keys?"

You back up. This is not the way you were planning to end the night. "Okay, look, I'll prove it to you." You whirl and trot to the house, with Cheryl close behind. Time to get some backup. If nothing else, you can muster up some peer pressure to bear on him.

You and Cheryl stop at the front door, looking back to where Walter stands, frozen, weaving a bit. Then you see him grin and reach into his back pocket. He pulls out his wallet, and from his wallet, he pulls what is apparently a spare key.

"I don't believe this," mutters Cheryl.

Laughing triumphantly, Walter slides into his car and starts it. You take a few steps toward the car, but you're not sure what you could do. The car whips around in the street, the forgotten beer bottle on the roof clattering onto the pavement, splashing beer. Walter peels out, hoping to prove just how in control he is.

You walk out into the street, Cheryl behind you. "You tried," she tells you. You bend down. The least you can do is pick up the discarded bottle before someone runs over it.

You hear the most incredible sound — like a pile of empty gasoline cans had been knocked over with a sledgehammer. Looking up, you see Walter's car, impossibly shortened, its front end wrapped around a tree like a pair of improvised pliers. As you watch, water from the destroyed radiator begins to pool around the rear wheels, looking like blood in the deepening dark.

You look to the house, where a number of people have come out the front door, summoned by the sound of the crash. You call to them, "Somebody call 911."

TURN TO PAGE 129 TO READ ABOUT PROFESSIONALISM.

# Cheryl: Invitation

To: Cheryl Stewart, MS3

Enrique Hernandez, MD

Cordially Invites All Health Science Center

Students

To His Annual

Incoming Class Party and Barbecue

Sunday, August 14

1:00 PM

1314 West 6th Avenue

Austin, Texas

Please RSVP. And, bring your swimsuit!

TURN TO PAGE 82, CHERYL, TO RSVP "YES."

# Cheryl: Barton Creek

YOU AND YOUR CLASSMATE, John Guerra, cruise along the rolling hills outside of Austin in your lemon-yellow, 1993 Honda Civic hatchback. It's a picture perfect day. The windows are wide open. The warm sun mixes with the sweet country air. Willie Nelson singing "On the Road Again" is cranked up loud. Today, there will be no lectures. No abdominal dissections. No latex gloves. The only agenda for today is to chill.

The GPS application on your iPhone signals your destination is 0.3 miles on the right. You turn at the sign that says BARTON CREEK and are flooded with fond memories. It wasn't that long ago that you would come up here on your day off to recharge and get centered. Your job as an emergency room nurse at Children's Medical Center was stressful. It was hard to forget your little patients when you got home. But, something about Barton Creek helped you cope. It truly is a magical place. You've only been back a few times since then, mostly for Dr. Hernandez' once-a-year party. It is absolutely gorgeous out here. The hills are covered in oak and cedar trees. Dr. Hernandez has a lovely home on a foothill that overlooks Barton Creek, a crystal-clear stream often referred to as "Austin's lifeblood" because it helps to supply the city with drinking water by replenishing the nearby aquifer.

It's 1:45 PM — just in time to be fashionably late. You pull into the driveway, shake John awake, put on your Ray-Bans and walk around the back of the house. The pool party is in full swing. A group of first-years huddle together around the hot tub. They look uncomfortable. You were uncomfortable once. John offers to get you a drink. "Beer? Wine? Vitamin water?"

You help yourself to a couple of towels and find a cozy spot by the pool for you and John to relax. There are many unfamiliar faces here today. It's a big school, but in many ways you would never know it. It's amazing how small your world becomes when you're in medical school.

John returns with your drink. You find a couple of chaise lounge chairs by the pool. Within minutes he is asleep again. You think about waking him so he won't miss the party, but you know how it feels, so you leave him be.

The water looks inviting and without hesitation you jump in. An inflatable mattress floats by and you grab it. You lie back and imagine that, someday, you will have a home just like this one. You realize your income depends on your particular specialty or subspecialty. Work as a county clinician and you can afford a moderate 2,100 square footer in a middle-class neighborhood with decent public schools, or specialize in implanting pacemakers and buy this house from Hernandez in ten years. It's still too soon to tell. Perhaps you'll marry another doc — *maybe a surgeon* — and combine your incomes so you can get the big house *and* still work for the county.

Your fantasy is interrupted by a piercing yell followed by a huge splash. Someone nearby has attempted a cannon ball turned belly flop. Your peaceful moment has been interrupted and you get out of the pool and dry off.

You land back in the lounge chair next to John with an audible sigh. He's awake now. "I don't know about you, but I needed this. Did you need this?"

John looks at you, then back at the pool. "I needed this."

"You know what I didn't need? The first year students keep getting younger."

John smiles, "It's not that they're getting younger, hon, it's that you're —"

There is an audible *snap!* as you bring your wet towel across John's chest. "I may be four years older than you, but finish that sentence and I will *finish* you."

"Fair enough," he grins. "Have you seen Dr. Hernandez yet?"

"Couple of times," you say. "He keeps coming in and out, playing host. Why? You looking forward to his annual big speech to the newbies?"

John drains his drink. "Oddly enough, yeah, I am. I think he really means all that stuff. At least, I hope he does."

"Oh, man," you say. "Remember *our* first year? When Dean Cutter gave the speech?"

John winces and pretends to spill his drink. "I started thinking about getting a job at my sister's dog grooming business after that train wreck."

"Yeah," you say. "I'm glad Enrique took that part over."

"Oh, so you're on a first name basis, now?"

You smile as you take a sip of your drink. "Nah. Someday. I'd like to be." You look around the pool, toward the large house overlooking the hillside. "I'd like a successful practice to afford a pool like this, too."

John rises from the chaise. "I think we need to get you graduated from med school first."

"Yeah, that would probably help."

He picks up his empty bottle and tosses it in a nearby bin. "My turn in the pool. Stand back. There's probably going to be steam when I hit the water."

You sit up in the chaise and towel your hair. Over your shoulder a shadow appears. Standing over you is a fairly attractive man, somewhat older than you. You think, *OMG* and sit up straight. The man takes the towel from your hand and, without saying a word, pulls your hair behind your ears. He then gently blots the water from your brow. You stare up into soft brown eyes, captured by a charming smile. He puts the towel back in your hands and walks towards the house.

*What was that all about,* you wonder? Perhaps it's time for another drink — something stronger this time.

You walk toward the house and find one of the many plastic coolers scattered about. Their contents are plentiful and varied. There's beer, of course, several local Austin brews. What else … soft drinks, bottled water, sodas, sparkling water … You make your decision and reach into the crushed ice, pulling out…

TURN TO PAGE 86 TO PULL OUT A BEER.

TURN TO PAGE 87 TO PULL OUT A SODA.

# Cheryl: Beer

YOU GRAB A COLD BEER and look for a spot to take in the view. John joins you, still dripping wet. You tell him about the tall, tan, stranger scene by the pool.

"Well, that's certainly weird." John grins with a look suggesting he doesn't quite believe your story.

You and John pause for a moment. Looking over the limestone patio you see a breathtaking panorama filled with lush, green foliage and an awesome view of downtown Austin. The two of you look out over the trees, clank you bottles together and say, "Cheers."

"Beautiful, isn't it?" a petite woman in stylish business attire asks.

Although you've never met her, you recognize the woman from the society pages. She is Dr. Hernandez' wife, Carla, a high-powered, well-respected trial attorney.

"I'm Carla Hernandez." She extends her hand. You shake it and introduce yourself. You thank her for hosting the party, and after a few moments of polite chitchat, she leaves. You return to the view when you hear someone yelling your name.

"Stewart! Guerra! Hey! Come over here, you two!"

TURN TO PAGE 88 TO CONTINUE.

# Cheryl: Soda

YOU GRAB A COLD SOFT DRINK and look for a spot to take in the view. John joins you, still dripping wet. You tell him about the tall, tan, stranger you just sort of "met" by the pool.

"Well, that's certainly weird." John grins with a look suggesting he doesn't quite believe your story.

You and John pause for a moment. Looking over the limestone patio you see a breathtaking panorama filled with lush, green foliage and an awesome view of downtown Austin. The two of you look out over the trees, clank your bottles together and say, "Cheers."

"Beautiful, isn't it?" a petite woman in styiish business attire asks.

Although you've never met her, you recognize the woman from the society pages. She is Dr. Hernandez' wife, Carla, a high-powered, well-respected trial attorney.

"I'm Carla Hernandez." She extends her hand. You shake it and introduce yourself. You thank her for hosting the party, and after a few moments of polite chitchat, she leaves. You return to the view when you hear someone yelling your name.

"Stewart! Guerra! Hey! Come over here, you two!"

TURN TO PAGE 88 TO CONTINUE.

# Cheryl: Photo Op

STANDING WAIST DEEP in the shallow end of the pool you see a fellow student. It's that first year, "Walter guy" you've heard someone — exactly who you can't seem to recall — talking about recently. *Something of a loose canon*, wasn't that the scuttlebutt? In one hand he has a ludicrously pink plastic camera. Walter is motioning to a group forming a few feet away in the pool. "I need a picture! Come on!"

You hang back, deciding if you should join them. The whole setup seems juvenile.

Walter motions everyone to get closer together. "Everybody got a beer? Good."

One of the girls raises her glass. "Rum and Coke okay?"

Walter grins. "Even better." He positions the group. "Okay, everybody, this is for my Flickr page. I need to show all my old frat brothers who are getting their MBAs that they made a poor life choice, so I need everyone to look like we are at the Party of the Century. It's my hope to make them want to slit their wrists with their spreadsheets." Walter looks up from the camera, a twinkle in his eyes. "So if any of you ladies want to show a little extra flesh, please feel free." He returns his attention to the camera. "Come on, Cheryl. Just because you're a third year doesn't mean you can't socialize with the peons."

TURN TO PAGE 89 TO JOIN THE GROUP PHOTOGRAPH.

TURN TO PAGE 94 TO DECLINE THE PHOTOGRAPHY SESSION.

# Cheryl: Cheese

YOU CLIMB BACK INTO THE WATER, press yourself into the mass of flesh and look at the camera. *Flash.* "Okay," says Walter, "you guys didn't look drunk enough. Try again, and get closer together."

There's some generalized splashing and you feel more bodies pressing around you as Walter begins to count down again. "Come on! That's got to be more enjoyable than you're making it look! One! … Two! … Three!"

The crush around you begins to relax, but Walter's still making like some clichéd tyrant film director. "More wanton! Think *Health Students Gone Wild!*" The crush returns around you. There are even a couple of calls of "Woo hoo!" Blue blobs are dancing before your eyes from the flash. Someone throws her arms around John. Someone else throws his arms around your waist and jokingly throws you into a back flip. *Flash.*

"Oh! Baby! That's what I am talkin about!" Walter is pleased with the photo session.

The pile of bodies pulls apart. Some head for the food. Some stay in the pool. You decide to head back to your chaise lounge.

TURN TO PAGE 90 TO CONTINUE.

# Cheryl: Flickr

TWO WEEKS LATER, you predictably have your hands full. You're certain there are people somewhere in the universe with busier schedules, but you can't find it in your heart to feel sorry for them.

There's something else, too. You're seeing a lot of knowing smiles and hearing an occasional, "Hey, Mama" or similar remark — like there's a secret organization of people around you that know something you don't. But, you also believe it's too early in the year for stress related paranoia, so you mentally shrug it off and try to bury yourself in your studies.

The comments keep coming, until one day you hear a, "Hey babe" from someone while you and John are visiting in the hallway. You ask him if he has any idea what that's about.

He looks at you pityingly. "You mean ... you don't know?"

You assure him you don't.

He pulls you over and begins setting up his laptop on a desk. As it boots, he asks you, "So you didn't get the email from Walter?"

Walter? You generally delete his emails. If he's not trying to set up a party, it's all whining about first-year stuff that doesn't concern you.

He angles the screen toward you. You instantly recognize the thumbnail photo layout of Flickr. You had an account back in the undergraduate days. You haven't paid much attention to it since ...

Wait a minute. You recognize some of those people in the picture. It takes you a second longer to recognize yourself. There you are. Someone with his arms around you. There's another with your right leg extended in mid air. And another with your left arm around someone's shoulder, your hand clasping a Dos Equis. You think, *OMG. No Wonder! Everyone thinks I'm a …*

Steam shooting from your ears, you head for the gross anatomy lab to find Walter. When you find him you are going to make him take out his laptop right then and there and make him take those photos off his site.

John rushes to catch up with you. "Oh, Cheryl, heh, heh, heh … Wait up! Let me show you what else a Google search for your name brings up …"

TURN TO PAGE 92 TO CONTINUE.

# Cheryl: The Letter

HAVING GIVEN WALTER A PIECE OF YOUR MIND you go back to your apartment. You check your mail and discover an envelope looking very innocent except for the bold, copperplate print of the return address: **OFFICE OF THE DEAN OF STUDENTS**. You think this can't be good. You pull it out and tear it open.

> Dear Ms. Stewart:
>
> It has been brought to our attention by the parent of a student that a certain series of photographs taken during a party at the beginning of the semester have been published on the Internet. These photographs document activities and behavior that do not reflect the image and reputation of our institution and its student body, of which we are justly proud.
>
> Therefore, this Office has no choice but to launch an investigation into these photographs and the people responsible. Please make an appointment with the Dean of Students as soon as possible.

You feel sick to your stomach. You know you should have never been in those damn pictures. And now, there's nothing you can do. You just hope the Dean won't come down on you too hard. *Yeah, right.*

TURN TO PAGE 88 TO TRY AGAIN.

# Cheryl: Camera Shy

YOU LAUGH AND QUIETLY WALK inside the house, away from the group, hoping the laugh will soothe any hurt feelings. A moment later, John rejoins you. "You didn't want in the group shot either?" he asks.

"Group shots I'll do. What Walter wants is an orgy."

There's more commotion in the pool. Walter is calling for a second shot. "More wanton! Think *Health Students Gone Wild!*"

"Well, why didn't *you* do it?" you ask John. "Not enough girls in there?"

He laughs. "Enough for me. Not for Walter, apparently. For me, it just didn't feel right. Probably something I would have done back in my freshman year, though."

You decide to tell John a little secret, a bit of personal history you've only shared with your mom and dad. It was back in your freshman year. You saw your roommate with a group of students hunched over a computer laughing at a website with photos of drunken girls. The photos had been taken off of personal sites and were plastered everywhere. You realized then that once stuff like that is out on the Internet, it stays out there forever. And if that stuff ever shows up — particularly after the word "Doctor" is in front of your name? "No way, Jose," you tell John. "I always want to be known as 'The Sober Girl On The End.'"

"Is this the wisdom that comes with age?"

You give him a piercing look and search for your wet-towel-whip when an older woman in a wildly colored sundress approaches you. Her wide-brimmed hat protects her lily white skin from the Texas sun. Her ringing call of "Hi-iiiiiii!" is like an opening salvo from an invading force.

"I'm Leslie Crenshaw, Dr. Crenshaw's wife? I'm so happy Dr. Hernandez asked us out here this year! We never seem to go out, it's all work, work, work, and study. You know how it is, right?" You and John inhale to answer, but she goes on. "This is a great party! Isn't this a great party? I really miss socializing like this. I mean, my book club is great, and they're all great people, but every now and then you've just got to go out and expand your horizons, am I right? I mean, you just stagnate, am I right? I'm really enjoying meeting all the other doctors out here!"

She stops abruptly, obviously waiting for one of you to reply. Caught in the glare of her thousand-watt smile, you try to think of something to say. Just as you open your mouth to speak, she continues, "So! What kind of doctors are you?"

Leslie smiles, eagerly awaiting your reply. Before you can answer, Walter walks by with a happy-go-lucky, "Hey, Cheryl."

You seize the moment. "Oh, ahh, hello! Um, excuse me Mrs. Crenshaw, it was nice to meet you," and turn to follow Walter. *That was convenient,* you think.

"Sorry you missed the photo session," he says as you catch up. "I suppose first year antics are a bit boring to an older woman like you."

You reply with clinched teeth. "You know, Walter. With age comes wisdom and skill. For example, I could probably make you pass out with simple pressure on your occipital nerves."

"Whoops, I didn't know you were so sensitive about age." Walter crosses his heart and raises his hand. "I hereby promise that I will never mention the 'O' word again."

You appreciate his apology. While you generally respond jokingly to someone's ribbing about being older than the rest of your class, it doesn't usually bother you — much. As you try to excuse yourself, Walter insists on showing you the photos. You laugh out of politeness but are thoroughly relieved that you did not participate in his *Animal House* session.

You try to excuse yourself again, but Walter keeps engaging you. You figure, *what the heck*, and chat a bit longer. During the conversation, Walter mentions how excited he is about the White Coat Ceremony at the end of the school year. You remember yours and assure him it will be one of the major highlights of his first year. Like a kid with a shiny new pencil box and fresh school supplies, he proudly tells you how he thoroughly researched and bought all of his supplies months ago. He is particularly proud of his Caribbean blue, stainless steel, Hittmann Cardiology III Stethoscope. But not because it can let you hear everything from the faintest heartbeat of an unborn child to a full grown adult. And not because it comes with non-chill rims and adjustable double-leaf binaural springs. And certainly not because it has soft-sealing ear tips or has his name engraved across the head. None of these are as important, Walter points out, as one simple feature — it was free.

For the first time during the conversation Walter has your full attention. Good stethoscopes are expensive and a Hittmann is considered top of the line. A wave of jealousy overcomes you as you think about your old, pink, off-brand, too-short, pediatric scope. You've had the thing ever since nursing school. It still works, but life would be sure better with a new one.

"That's a nice present, Walter," you gotta admit.

"Oh, it wasn't a present." Walter thinks for a moment before continuing. "Hey, I bet you could get one too."

"I could? How?"

Walter explains that the scope was given to him by a pharmaceutical representative. He met the rep at a conference. The rep said that her company likes to help out medical students, so they authorized her to provide students with free textbooks or basic instruments. They have access to Hittmann's discontinued medical devices. Walter produced a card from his wallet with the rep's information.

"Here, Cheryl. Take this card. I bet she can get you one too."

You hold the card in your hand. Wow, you think, a new stethoscope. A good stethoscope. You always look at them online and fantasize about having one with long tubing. It's terribly distracting to be looking up your patient's nose every time you listen for a heartbeat. And it would be great to have one in a conservative color. The pink was okay for pediatrics, but a brand new, twenty-seven inch, in simple gray? ... *Yes, very professional.*

You notice the rep's mobile number is on the card. You could probably leave a voicemail and get together sometime during the week. You reach into your bag and feel your cell phone lying at the bottom. *Or should I?* You realize that accepting something for free, even a cup of coffee, from a sales rep of any kind is ethically complicated. You start thinking, *what's the harm in taking something that's discontinued?* If I don't take it, it will probably be shipped off to some third-world country to be used by some Doctor Without Borders. I don't mean to say it isn't important for poor countries to get supplies, but they usually get help from charitable organizations … and I would be taking just *one*. Who knows, maybe I'll even volunteer myself one day and they wouldn't have to furnish me with a stethoscope, because I would already have one. *Right?* You reach in your bag and pull out your phone.

TURN TO PAGE 99 TO CALL THE DRUG REPRESENTATIVE.

TURN TO PAGE 105 TO KEEP YOUR OLD STETHOSCOPE.

# Cheryl: Heather

THE DAY AFTER THE PARTY, you get back to business. As you roll out of bed you remember the business card Walter gave you. It's 5 o'clock in the morning and you fumble for your bag near the nightstand. Your phone falls out, face up, and turns on. The screen reads, "MISSED CALL: HEATHER MONROE."

You don't know a "Heather Monroe," but as you reach down from the bed, you spot something else that tumbled out with your phone: a business card, lying face down on the floor. You reach down and retrieve the phone and card. You're about to delete the missed call message when you notice the card reads, "Heather Monroe, Sales Representative." *Oh yeah, now I remember.*

You push REDIAL on the phone. While you don't expect the sales rep to answer at this ungodly hour, you want to leave a message before you get on with your day. To your surprise your phone rings less than sixty seconds after you've left a voice mail.

"Hi Cheryl, thanks for calling."

"I am so sorry to wake you, Ms. Monroe, I —"

"No, no, you didn't wake me. I'm up and having coffee already. I'm an early riser. It's going to be a beautiful day! How can I help?"

*Nobody has the right to be this perky in the morning*, you think. Must be genes. *Or incredibly strong coffee.* You explain how you received her card from Walter and that you didn't know if she had any more stethoscopes, but that you could sure use one. You throw in that you understand these are discontinued models and not considered of value anymore. Heather corrects you that these are brand new. She informs you that not only does she have one, but she will make sure you get exactly what you want — color and all — and with your name engraved across the chest piece. Heather offers to meet you by the end of the week to deliver it to you, personally. You and Heather arrange a meeting by the sushi bar in the campus dining facility. She invites you to bring any of your classmates who might also be in need of a new scope. You agree and say that you will ask around to see if anyone else would like to come along. You hang up, excited. *Cool,* you think, *a new scope.*

The second alarm clock sounds (you have four scattered around the apartment to make sure you get up). You make a quick cup of coffee and grab a protein bar as you head out the door for another fun-filled, jammed-pack day as a med student.

TURN TO PAGE 101 TO CONTINUE.

# Cheryl: Gray Scopes

YOU'RE IN THE CAFETERIA LINE after a grueling morning of rounds that was followed by an obstetrics lecture, which was capped by a long observation of a cesarean that produced three beautiful baby boys. Lunch is going to taste good, even this food. You spy John three persons ahead of you in line, who turns to see you almost simultaneously. John lifts his tray and walks back to join you.

"Hey," he says.

"Hi," you reply. "Buy you lunch?"

"Why, indeed you may! I'm quite found of their Barons Rothschild 1954 and chateaubriand special." John brings his fingers to his lips and blows a kiss. "Impeccable."

"Settle for a tuna sandwich and soft drink combo like mine?"

"That was my second choice."

The two of you chat. John is excited because he has decided to run for president of the local AMSA chapter. In your mind, John would make a perfect president for the student association. You congratulate him on his decision and offer to help with his campaign.

As he thanks you, you notice John's eyes are not looking at you. John is staring at your neck. At least, you hope it's your neck he staring at. "Man, that's *sweet*, Cheryl." He lifts the chest piece of your shiny new Hittmann stethoscope, which you can now wear around your neck because of its extra long tubing. "Top quality." He turns it over. "Engraved! How did you manage to spring for it?"

You tell him about the sales rep and offer to pass on her business card so he can get a stethoscope too.

John's face turns sour. The scope drops back on your chest. "Cheryl. I am surprised at you."

You're not sure if he's joking or serious. Then you figure he's serious but are not sure why he's upset. "What for?"

"*What for?* That rep is bribing you, that's *what for*. She is trying to get you into the habit of using, and then eventually buying, her products. Didn't you learn anything in ethics? Does, *There's no such thing as a free lunch* ring a bell?"

You get a little indignant. "Get real, John. It will be many years before I'm able to prescribe anything." You want to say, *for goodness sake, John, get off your high horse*, but you don't. Instead, this comes out: "You think I'm not smart enough to know when I'm being manipulated? Give me a little credit, please. There are no strings attached to this gift, not in my mind, at least. The Hittmann rep is being gracious, that's all. When one day, in the far distant future, I prescribe meds, I may remember her kindness, but nothing more. I'll prescribe the right medicine for my patient, and it will be whatever provides the most value to my patients, not the most gifts to me. You dig?"

John says nothing.

"Okay, I know what you're thinking. It's a slippery slope and blah, blah, blah ..."

John says nothing.

"Oh, and I'll bet *you* never took anything from a sales rep? A cheap pen? A notepad? Are you telling me you never attended an education seminar they funded? Or ate the lunch they paid for? Not even the donuts they bring to grand rounds? *Come on!*"

John says, "Anywhere But New Jersey, eh?" then turns toward the front of the line. The cashier is just coming up.

"What's New Jersey got to do with anything?"

John peers over his shoulder. "A-B-N-J. You're a big girl. Look it up."

Okay, now you are mad. "Well, you're right about there being no free lunch, John. Buy it yourself." He does.

The two of you find different tables at opposite ends of the cafeteria. You both sit alone.

You're replaying the conversation over and over in your head. *What just happened?* You think about moving to John's table. He hasn't had the pleasure of hearing your full explanation yet. Like, why you'll be able to provide better care to your patients because you have better equipment. Or ... but as you think, you realize John *may* be right. Could you be rationalizing the "gift" because you wanted and needed it so badly? Yeah, John is right. You gotta give it back, or something. You have to maintain your integrity, no matter what.

You take the new scope off your neck. While holding it in your hand you cycle through Ross' Stages of Grief, going all the way from shock to acceptance in less than sixty seconds. It's hard as hell to let go of your "gift," but you do. You make a personal vow to send your new stethoscope to Doctors Without Borders tomorrow. Tomorrow morning, in fact, first thing — before you have a chance to regain your sanity.

You are sure you are doing the right thing, but somehow the "right thing" doesn't feel so good. You liked that damn scope. You toss the remainder of your tuna sandwich on the tray and head out. With grief comes loss of appetite. It just turned 1:00 PM. Time to get back to the hospital and check on the patients who are about to deliver. You figure this will be the last time you use your new Hittmann to listen to the tiny *lub-dupps* of the newly born. *I hope some guy in Uganda appreciates this.*

TURN TO PAGE 94 TO TRY AGAIN.

# Cheryl: Pink Scopes

YOU DROP THE PHONE BACK INTO YOUR BAG. *Nah, my old stethoscope's got personality,* you say to yourself. *I'll buy my own scope when the time is right, and with no strings attached.*

You head back to where you left John … he's still there, standing by the pool and looking a little forlorn. Mrs. Crenshaw's mating call can be heard somewhere off in the distance. As you approach John, you hear a voice behind you say, "Excuse me!" You both turn to see a young man with dark, wavy hair, and skin even darker than your own approaching. You recognize him as one of the students in Walter's group shot, but that's about it. "Where do you guys know Mrs. Crenshaw from?"

You explain that you just met her.

He introduces himself, "My name's Parvesh. In the dental program? The Crenshaw Center?"

You remember, *Oh, yeah! Dr. Crenshaw. Dental!*

Parvesh nods. "Dr. Crenshaw told some of the dental students about the party, said there would be a practicing dentist or two here, if we had any questions, you know?"

You tell him that Dr. Hernandez works with many of the local doctors. You introduce yourself. Then you introduce John as a Neanderthal you found in some sage brush along the road on the way to the party. You assume John understands this is a payback for the older woman comment he made earlier.

"Pleased to meet you both," says Parvesh.

Parvesh is curious about everyone. He points to a bar area where a statuesque redhead, wearing a large blue cambric work shirt open over a modest black bathing suit, is carrying on a conversation with Dr. Hernandez. You recognize the woman in the movie star oversized sunglasses as Dr. Frey. She's the new university research integrity officer, and has an informatics degree. You also see your friend Julie Spates there. Like you, she was an ER nurse in a previous life. Now she's back in school to become a nurse practitioner. But, you figure, Parvesh is probably not asking about her.

John chimes in. "Margeaux Frey, right?"

You pat him on the head and say, "Good Neanderthal, good. Moving up the evolution chain now."

Parvesh smiles and then looks furtive, as if telling a secret. "Have you heard her French accent?"

You don't respond to his comment. Instead, you say, "And who's *that?*" pointing to the tall, dark, handsome stranger who likes to dry women's hair. He's just joined the Frey, Spates and Hernandez group with a bottle of wine in each hand, and is apparently asking the other three's opinions.

"That's Dr. Cochran," John blurts. "Clinical faculty. Surgery. I think he runs the surgical residency program."

*Hmm,* you think. *Very interesting.* "I'm thinking about a surgical residency."

Another gentleman walks into Cochran's group. Older and more distinguished in both dress and demeanor, and puts his arm around Leslie Crenshaw. "Oh, there's Dr. Crenshaw," Parvesh notes. It looks like Crenshaw and Cochran are good friends. Each slaps the other on the shoulder in gentle laughter.

You notice that a decision was apparently rendered in the doctors' wine conference, as Cochran walks into the house, perhaps looking for a corkscrew or something. John notices you watching intently. He's also figured out Cochran's the guy who dried your hair. John leans toward you whispering, "and I don't think he has a French accent, Cheryl."

"I'm going to leave you two boys to your fantasies about French women. Me thinks it's time for a refill."

TURN TO PAGE 108 TO CONTINUE.

# Cheryl: Refill

YOU WALK OVER TO THE HOUSE. Cochran has disappeared and in his place is a woman wearing a headscarf. She's standing, talking with Leslie Crenshaw and a man with his back turned to you. He turns around and you instantly recognize Lawrence Crenshaw, Director of the school's Crenshaw Center.

The woman in the headscarf is looking a little uncomfortable. You think she must be terribly hot in this weather. Hernandez has the air-conditioning on but still, a headscarf in Austin in August? You walk over just as the Crenshaws are leaving and their conversation winding down.

"It was good to meet you Dr. Crenshaw, and you too, Mrs. Crenshaw," the headscarf woman says. She and Dr. Crenshaw shake hands. "Thank you very much for the offer," she tells him. "And I will certainly call your office next week and arrange that appointment." The Crenshaws disappear into the crowd, leaving the headscarf woman looking a little lonely.

Just as you approach her, Dr. Cochran returns. *Whoa,* you think. *Here he is again.*

Cochran smiles charmingly at you. "Well, hello there. I'm John Cochran." He extends his hand. "Who do I have the pleasure of meeting?"

"I'm Cheryl ... um ... that is Cheryl, in ... I'm a medical student? ... in ..." You can't believe your stammering. *Why am I so nervous?*

Cochran takes your outstretched arm and clasps it with both hands. "I am delighted to meet you, Cheryl." He looks directly into your eyes, which makes you even more nervous, then turns to the woman in the headscarf. "Faiza, I'd like to introduce you to my new friend, Cheryl The Medical Student. Cheryl? Meet Faiza The Researcher. Faiza? Cheryl." Cochran is still holding your hand.

You smile awkwardly at Faiza. She rolls her eyes and shakes her head at Cochran, then extends her hand to you. "I am *Dr.* Patel and I am glad to meet you, too, Cheryl."

Cochran finally lets go of your hand. "So, may I get you ladies something to drink?"

"Yes, a beer. Um, a beer would be lovely. Thank you." You are still stammering.

"And how about you, Faiza?" Cochran asks. "Will you join us in a beer? Maybe some wine?"

"You know I don't drink, Dr. Cochran. It is against my religion."

"Oh, I know, I know — but come on! You're among friends here, Faiza. You can let your hair down occasionally, can't you?" You can tell that he's taunting Dr. Patel, though smiling as he speaks. "Just look at everyone outside by the pool enjoying themselves. Have you ever even been *in* a bathing suit?"

Your fantasy image of Cochran slowly dissolves. *So much for being a gentleman.* There goes the double income, large house and matching Porsches. You begin to think that maybe you had better not accept the beer. You don't want Dr. Patel to feel any more uncomfortable than Cochran is obviously trying to make her. Besides, Dr. Cochran is starting to make *you* a little uncomfortable. You're not sure if it's just authority figure jitters or something else.

"You know Faiza," Dr. Cochran continues, "there's a famous book you might be interested in." Cochran winks at you. "I'll give you my copy next week. It's called *Our Bodies Ourselves*."

"Thank you very much, Dr. Cochran. I should like to give you a famous book as well, a copy of the Qur'an." Dr. Patel holds her decorum.

You can't believe what you just heard. While you agree that *Our Bodies Ourselves* is a good book — your mother gave you the one she read when she was in college — you are sure Dr. Cochran is using it as some kind of dig. *How can he be so insensitive?*

Their conversation starts to get a bit more animated, and not in a good way. You decide to do something to interrupt the conversation.

TURN TO PAGE 111 TO TALK TO PATEL.
TURN TO PAGE 114 TO TALK TO COCHRAN.

# Cheryl: Patel

"THAT'S A BEAUTIFUL HEADSCARF, DR. PATEL." Your voice is steady and composed.

"Thank you. It is Indian silk."

Success. The two of you are consciously ignoring Cochran. "Oh, is that where you're from? I have an ophthalmologist friend who just returned from a few weeks in Madurai and had a wonderful time." Cochran walks away, heading toward a beer cooler.

"Ah. Probably the Aravind Eye Hospital. It is a very famous place. But, that's in Chennai, or what used be called the state of Madras. I am from the state of Gujarat, entirely on the other side of the country."

Cochran is now safely out of range, but nevertheless, you speak softly. "I'm sorry he offended you."

Faiza is silent for a moment, then speaks. "Westerners don't understand my culture. They always want to paint the world with a broad brush. Often times, though, it is the wrong brush. *My* headscarf is *my* choice. *My* career is *my* choice. If I thought I was being oppressed I would certainly do something about it. I don't need a chauvinistic misogynist to tell me what I should and should not do."

You agree with her. You feel as if you should apologize for all Americans. You reassure Dr. Patel that not everyone feels that way and, while you can't guarantee that it won't happen again, you tell her that if she ever wants to talk about it, or talk about anything, she can count on you.

She places a comforting hand on your shoulder and looks you in the eye. You exchange no words, yet speak volumes. "Oh, this has happened before. Last year, I took my mother to the emergency room for chest pains. Thank goodness it was not a heart attack, but we had a terrible time in the ER. They had no idea of her modesty, especially the male doctors. They quite honestly didn't know how to handle the situation, how to examine her."

You've heard the ER rotation can be stressful in many ways. You make a mental note to ask Faiza more about what happened with her mother when you notice Parvesh standing close by. *Perhaps these two should know each other? It might be comforting for Patel to talk to a compatriot.* You motion Parvesh over. "Faiza, let me introduce you to Parvesh … um …"

"*Singh*. It's Parvesh Singh." Parvesh extends his hand.

"Right. Let me introduce you to Parvesh *Singh*."

Patel shakes his hand. "I am Dr. Patel. Pleased to meet you, Parvesh. Are you a medical student too?"

"I am a dental student. And you?"

"Epidemiology researcher."

"Oh, I am interested in research too."

"Well, you just missed Dr. Crenshaw. He invited me to collaborate in a longitudinal study he's running on a new drug for oral cancer. Perhaps you would be interested? I know he's looking for more help."

"That would be wonderful, yes."

"I have a meeting with him next week. We can both attend if you're available."

Cochran returns with your beer. Before he can speak, Hernandez comes in from the pool and asks everyone to come outside for a few minutes.

TURN TO PAGE 116 TO HEAD OUT TO THE POOL.

# Cheryl: Cochran

"SO DR. COCHRAN, HOW BIG is the general surgery residency?" Faiza walks away. You feel bad for her, but Cochran doesn't seem to notice the absence. He is intently focused on you.

"We take ten residents a year. It's a wonderful program. Are you interested in surgery, Cheryl?"

"Um, well, yes, but ..." *Okay, here goes* ... "Um, Dr. Cochran ... I hope you don't mind me saying this, but —"

He interrupts you with a raised hand. "Yes, yes, I know. Who am I to say the woman needs to loosen up and join the twenty-first century? Is that what you were going to say?" Cochran has turned off the charm.

"Something like that, yes sir."

"I'm doing her a favor, Cheryl. This is the United States. She's in the United States now. She studied here and now she is working in the United States, not living and working in India. I'm a firm believer in the saying, 'When in Rome, do as the Romans do.'"

"Yes, that's one perspective, sir. There is also the belief that cultural diversity has its own benefits."

"Of course, of course." The lights in Cochran's eyes snap back on. "And there is *my* belief that hot tubs also offer great benefits." His smile turns impish. "Care to join me?" He gestures toward the hot tub just outside the living room window.

*Hmm.* You're not sure where this conversation with Dr. Cochran is going, but it sure has been interesting. In particular, you wonder how someone who can be so charming and obviously smart can also be such a hardhead. You wonder if it's a function of his age or his politics that makes him close minded. Your contemplation is interrupted as Hernandez comes in from the pool. He asks everyone to come outside for just a few minutes.

TURN TO PAGE 116 TO HEAD OUT TO THE POOL.

# Cheryl: Après Vous

DR. COCHRAN OPENS THE DOOR for you as you step outside. A fourth year student near the house starts knocking two empty wine bottles together; careful not to break them, but producing a clear, remarkably loud ringing sound. Nearby, Dr. Hernandez is climbing a stepladder brought from the house to give him some extra height. The music and general hubbub quiets as more and more heads turn toward the doctor, who stands with his arms outstretched, motioning for quiet.

"If I could have your attention for just a minute, please," he calls out, his voice strong and loud without shouting. "I have a few things I'd like to say."

The party dampens even more, a few people strolling toward the house to get a better view. The bathers empty the pool, towel off, and head into the crowd. "First of all," he continues, "I'd like to say we are very pleased with this wonderful new class, chosen so carefully by our Admissions Office."

There is a boisterous, "YEAH!" from the back of the crowd. A quick glance proves the source to be Walter waving a beer in his hand.

"And so enthusiastic," smiles Hernandez. A general chuckle runs through the crowd.

Hernandez goes on, "Your application essays were read, your interviews were considered, and it was felt that each of you had the makings of a good healthcare professional. You were all tops in your class, but what is more important, you all had the sort of character we want to be associated with our school."

You glance quickly back at Walter, but he, like the rest, seems to be listening raptly.

"Your essays told us of a deep motivation to relieve suffering, to research and cure disease, to be a positive force in the world. Reading those thoughts makes me proud of my profession, and proud of you, who will become my colleagues."

Hernandez changes his stance slightly. "You've got something of an adventure ahead of you. When you applied to this school, you weren't just trying to get into a highly competitive graduate educational program. No, it was like you were proposing marriage to someone."

There is a general wave of laughter through the crowd. Hernandez holds up a hand. "No, no, hear me out. When a school interviews you, it's like a first date. They're finding out what kind of person you are, if they want to be seen with you in various settings. When we admitted you, we accepted your proposal. We didn't just like your grades, we liked *you*.

"Now, once you were admitted, you still had the chance to change your mind. Some people get cold feet when they're faced with the cost and complexity of a wedding. But if you don't cancel, if you proceed, you are making a lifetime commitment to another person. You care what they think about you, and you're willing to make some personal changes to meet their expectations. Becoming a healthcare professional is not just getting a degree, or learning a lot of new information. It also means a change in your identity. First, others will start to see you differently. Later, you will start to think differently, and finally, you're going to see yourself differently as well."

He gestures to the crowd at large. "Most of you here are entering medical school students. I want to point out that we've also invited some of our more advanced students, up-and-coming second and third year folks, so you can get to know them, and maybe even be gifted with some wisdom about the weeks and months to come."

He makes a show of looking over the faces before him, pointing to various people. "I also want to acknowledge that we've got some colleagues with us, students and faculty from the Dental School, Nursing School, the Graduate School and … there you are! The School of Public Health! I'm glad all of you could make it!"

You survey the unfamiliar faces in the crowd.

"So, you may be asking yourself, 'Why the eclectic mix?'" Dr. Hernandez smiles in a dramatic pause. "Because I wanted to underscore a fundamental truth about our health professions. All of us — physicians, nurses, dentists, researchers, public health and bioinformatics professionals — are part of the same team."

Hernandez squints as he surveys the crowd. "I also see a few of my patients here today too, but don't worry, I won't identify you. HIPAA, you know." A smattering of laughter follows.

Hernandez glances down at his feet, gathering his thoughts, then looks back up. "If there is anything you take from me standing up here and shouting out at you, it's this: It is important to remember that we, all of us, every school, every discipline, share a common goal: to improve the quality of healthcare, and its delivery. And, it is equally important to realize that none of us can do that alone. We can only be effective as a team. So please say, 'hi' to someone you don't know today. Get to know your team. Again, welcome."

Hernandez steps down from the ladder to no small applause. You find yourself nodding appreciatively. That actually *was* better than last year's.

There is a tap on your shoulder. Cochran, still smiling, extends his arm toward the hot tub. "*Après vous?*"

*Oh well, what the heck.*

* * * *

The relaxing hot waters feel great. There are six of you in the tub as the sun slowly sets. As the party winds down, the music and conversation become more subdued. You are just about to fall asleep when you notice the tub is now empty save for you and Dr. Cochran. And, he is sitting close, very close, and getting closer.

Cochran reaches an arm around your shoulder and leans into you. You instinctively laugh as you try to organize your thoughts. *Okay, should I say something? Move away from him? Slap him?* On the other hand, that could kill any chance of getting a surgical residency here, maybe even anywhere. If you reject him, *then what?* Cochran plants a small kiss on your cheek. *Decision time.*

TURN TO PAGE 120 TO KISS COCHRAN BACK.
TURN TO PAGE 123 TO PUSH COCHRAN BACK.

# Cheryl: The Kiss

YOU DECIDE A SMALL KISS on the cheek is harmless, friendly even, and lean over to return the favor. As you sit back, you notice Dr. Hernandez in your peripheral vision standing just inside the living room sliding glass door. *Did he see that? Is he watching us?* The awkwardness is broken by a nearby splash.

It's John. "Come on Cheryl. It's time to go home and feed the piranhas. They haven't been let out all day."

You slowly untangle yourself, climb out of the hot tub, and wave bye-bye to the good doctor. Your body shakes in the cool air as John hands you a towel. The two of you decide that now would be an excellent time to leave the party.

TURN TO PAGE 121 TO CONTINUE.

# Cheryl: Reprimand

YOU ARE AT THE FRONT DOOR to say your goodbyes to Dr. and Mrs. Hernandez. There's something of a reception line forming. Everyone wants to thank their hosts on the way out.

Finally, it's your turn. "As usual, this was a great party, Dr. Hernandez. It's so wonderful that you open your home up every year to us."

He smiles in a strange way, than gently tugs your elbow. "Cheryl, can I have a word with you?" He walks you down the drive, just out of range of the crowd at the door. You feel uncomfortable as he speaks. "Look, Cheryl ... I ..." He clears his throat. "You may not think this is any of my business, but I am going to make it my business."

You think you know where he's going and open your mouth to speak, but Hernandez raises his hand.

"Please hear me out. Whatever is going on between you and Dr. Cochran stopped five minutes ago. Whatever that was, it's over. Do you understand?"

You nod your head.

"I know you, Cheryl, and I know you are smarter than that. I also know that *you* know there is no such thing as a consensual sexual relationship between a professor and a student. It's highly unethical. The presumption will always be that someone was coerced, whether it was intentional or not."

"Yes, sir. I mean, no sir. I mean ... really, there is nothing between —"

Again, he cuts you off. "Good. Let's keep it that way. You wouldn't pursue a sexual relationship with a patient, would you?"

You are shocked by the analogy. The thought of anyone breaching the doctor-patient relationship makes you sick.

Dr. Hernandez answers his question for you. "Of course you wouldn't. The same rules apply in this situation. Have I made myself clear?"

You assure him you understand everything he has said and that it will never happen again. You feel horribly embarrassed and grateful at the same time. You catch up with John who has that *what the hell happened* look on his face. As you walk toward your car you fill him in on all the horrid details.

TURN TO PAGE 116 TO TRY AGAIN.

# Cheryl: Push Back

YOU GENTLY push the good doctor back with your hand. He smiles in a "just kidding" kind of way. The tension is broken by a nearby splash.

It's John. "Come on Cheryl. It's time to go home and feed the piranhas. They haven't been let out all day."

You slowly untangle yourself, climb out of the hot tub and wave bye-bye.

"Thank you," you confide to John, your body shaking in the cool air. He gives you a reassuring pat on the shoulder and a towel. The two of you decide to leave the party before anything else weird happens.

TURN TO PAGE 124 TO CONTINUE.

# Cheryl: Drunk Walter

DRY AND DRESSED, you and John say so long to the new students you've met throughout the day — the ones still hanging around, at least — and thank your host. As you both walk out the front door, the first thing you notice is Walter standing in the street, next to what is apparently his car. He is pawing through his duffle bag, and an open bottle of beer sits on the car's roof. "Oh," you say, "this is going to get very sad very quickly."

You and John approach him as his search through the contents of the duffle bag becomes more frustrated. "Hey, Walter," you call out to him. "What are you doing?"

He pauses in his rummaging to look up at you. "Oh, hey." He starts patting on the pockets on the side of the bag. "I'm jus' lookin' for the keys to my chariot, y'know? I mean, I gotta have 'em. I got here, right?" He suddenly puts his hand in his pants pockets and laughs, a sputtering giggle. He pulls the keys from his pocket. "Las' place you always look, am I right?" He attempts to sort through the keys and instead drops them on the lawn. "Aw, crud."

As you watch Walter laboriously bend over to retrieve his keys, you realize you're not quite able to retreat to the simplicity of your apartment yet. John had said something earlier about Walter binge-drinking all day. There's nothing wrong with having a drink every now and then to relax, but Walter seems determined to burn out his liver before school even starts. Walter might bear a bit of watching in the coming year — his current state might be a warning of future problems, maybe even a continuing problem the Dean should know about.

Of course, you and John have the problem right now. "Walter," you say, "I'm not sure you should be driving."

"Nonsense!" Walter shouts. "Drivin's like autopilot to me. I'll be fine …" and in the course of gesturing, his keys go flying into the lawn. "Aw, crud," he repeats.

TURN TO PAGE 126 TO TAKE WALTER'S KEYS.

TURN TO PAGE 128 TO LEAVE WALTER TO HIS OWN DEVICES.

# Cheryl: Walter's Keys

JOHN HAS A SLIGHT ADVANTAGE OVER WALTER. He snatches up his keys from the grass. Walter holds his hand out for them. "I'm serious, Walter. You can't drive," John tells him.

"Yes, I can," he slurs, trying to sound sober. "Just gimme my keys."

"No. Either Cheryl and I take you home, or we call you a cab. That's your choice."

Walter advances on John, and he's starting to look menacing. "Will you jus' gimme my dang keys?"

You and John back up. This is not the way you were planning to end the night.

John says, "Okay, look, I'll prove it to you." He whirls to trot to the house. You take off after him. *John feels a need for more backup*, you figure. If nothing else, maybe a group of you can muster up some peer pressure to bear on him.

You and John stop at the front door, looking back to where Walter stands, frozen, weaving a bit. Then you see him grin and reach into his back pocket. He pulls out his wallet, and from his wallet, he pulls what is apparently a spare key.

"I don't believe this," you say.

Laughing triumphantly, Walter slides into his car and starts it. You take a few steps toward the car, but you're not sure what you could do. The car whips around in the street, the forgotten beer bottle on the roof clattering onto the pavement, splashing beer. Walter peels out, hoping to prove just how in control he is.

You walk out into the street, John behind you. "We tried," you say. As you bend down to pick up the bottle, you hear the most incredible sound — like a pile of empty gasoline cans had been knocked over with a sledgehammer. Looking up, you see Walter's car, impossibly shortened, its front end wrapped around a tree like a pair of improvised pliers. As you watch, water from the destroyed radiator begins to pool around the rear wheels, looking like blood in the deepening dark.

You look to the house, where a number of people have come out the front door summoned by the sound of the crash. You call to them, "Somebody call 911!"

TURN TO PAGE 129 TO READ ABOUT PROFESSIONALISM.

# Let Walter Drive

YOU HEAR THE MOST INCREDIBLE SOUND — like a pile of empty gasoline cans had been knocked over with a sledgehammer. Looking up, you see Walter's car, impossibly shortened, its front end wrapped around a tree like a pair of improvised pliers. As you watch, water from the destroyed radiator begins to pool around the rear wheels, looking like blood in the deepening dark.

You look to the house, where a number of people have come out the front door as if summoned by the sound of the crash. You call to them, "Somebody call 911!"

TURN TO PAGE 129 TO READ ABOUT PROFESSIONALISM.

# 1: Professionalism

# 1.1: Definition of a Profession

YOU MAY BE WONDERING what is meant by a "profession." Is it just a word we could apply to anything to make it seem important? Sociologists and historians have studied this. A profession is a socially accepted practice that involves three basic elements:

- Expertise
- Self regulation
- Fiduciary responsibility

*Expertise* means a high level of education in a field requiring knowledge and/or skill under the guidance of others who supervise one's acquisition of knowledge during a period of practice. In the health professions, such expertise may involve years of classroom, laboratory, and internship training devoted to acquiring both technical knowledge and practical application. To qualify as a profession, a field must have a specialized body of knowledge that requires mastery.

*Self regulation* means that members of the profession are given authority to set standards and police themselves. In clinical health professions, one must pass a licensing exam and submit to the oversight of state regulatory boards staffed by people with specialized disciplinary knowledge of the profession's ethical norms. Standards typically include knowledge, skills, and ethical behavior. Violation of a profession's standards may result in expulsion. Maintaining these standards requires members to report those who fail to meet the standards. In healthcare, substandard behavior constitutes a danger to patients or the public.

*Fiduciary responsibility* means that members of the profession are obliged to put the welfare of those they serve ahead of their own interests. That is, a profession requires altruistic motivation and behavior, distinguishing it from being a purely commercial enterprise.

# 1.2: Codes of Ethics

MOST PROFESSIONS HAVE A WRITTEN CODE OF ETHICS.
Codes are usually written by a panel of leading members of a
profession. Some codes focus on establishing minimally
acceptable behaviors, violation of which would mean
expulsion from the group and losing your license. Other codes
are more idealistic or aspirational. Large, well-established
professional fields often have multiple codes of ethics. Nursing
has a code of ethics written by the American Nurses
Association, just as dentistry and public health have their own
codes. Medicine has multiple codes.

Every profession is ethically grounded in fiduciary
responsibility, and each profession specifies that responsibility
according to its own area of expertise. Review the code of
ethics for the profession you are entering below, then explore
one other code. Compare the two and identify the important
areas of overlap. In what ways do they differ?

- *American Dental Association, Principles and Code of Ethics* (http://
  www.ada.org/sections/about/pdfs/code_of_ethics_2011.pdf)

- *American Medical Association, Principles of Medical Ethics, 2001*
  (http://www.ama-assn.org/ama/pub/category/2512.html)

- *American Medical Informatics Association, Code of Professional
  Ethical Conduct* (http://www.amia.org/about-amia/ethics/
  code-ethics)

- *American Nurses Association, Code of Ethics for Nurses* (http://www.nursingworld.org/MainMenuCategories/EthicsStandards/CodeofEthicsforNurses)

- *American College of Epidemiology, Ethics Guidelines* (http://www.acepidemiology.org/statement/ethics-guidelines)

- *Non-governmental Organizations, Code of Conduct for Health Systems Strengthening* (http://www.ngocodeofconduct.org)

- *Principles of the Ethical Practice of Public Health* (http://www.apha.org/NR/rdonlyres/1CED3CEA-287E-4185-9CBD-BD405FC60856/0/ethicsbrochure.pdf)

- *American Health Information Management Association, Code of Ethics* (http://www.ahima.org/about/ethicscode.aspx)

- *American Society for Microbiology, Code of Ethics* (http://www.asm.org/ccLibraryFiles/FILENAME/000000001596/ASMCodeofEthics05.pdf). The American Society for Microbiology website is a good example of codes for scientific fields. You can also look for comparable codes in your field of biomedical science, and look under the National Science Foundation, the Institute of Medicine, the National Academy of Sciences, and the American Association for Advancement of Science.

- *American College of Physicians* (http://www.acponline.org/running_practice/ethics/)

# 1.3 Interprofessional Ethics

AN UNUSUAL AND IMPORTANT ASPECT of ethics is that it is *transdisciplinary*. We don't seek different ethical qualities in dentists and nurses, for example, but encourage them to have shared values to work together collaboratively. The same can be said for physicians, dental hygienists, researchers, public health officials, and other clinicians. In each case, your fiduciary responsibility is primary. The goal of interprofessional ethics is to identify and resolve ethical conflicts collaboratively.

*The Brewsters* provides a basic introduction to ethics that are the same across schools. All students will have a common vocabulary for identifying and resolving ethical issues, and for respectfully settling differences between members of a healthcare team. All professionals should demonstrate humility about their professional role while engaging collaboratively with other healthcare professionals, acknowledging the contributions and expertise of others and actively soliciting others' opinions.

# 1.4: Misleading Titles

THE IMPORTANCE OF TRUTH TELLING, keeping your ego in check, and putting patients' interests ahead of your own might be best seen in the simple matter of how you introduce yourself (or allow yourself to be introduced). If you are a dental, medical, or nursing student, you should be introduced as such. "Doctor" is not accurate until you have graduated with a doctoral level degree (and some would go further and say not until you have passed the Boards, if your field has them). Graduate students are not allowed to publish papers with "PhD" after their name until completing their dissertation defense for similar reasons, and graduate schools can be especially harsh towards students who represent themselves as having a PhD already in hand.

Being called "Doctor" before you complete the MD, PhD, DNP, or DDS degree is deceptive and potentially harmful. In Act 1 of *The Brewsters*, medical student John Guerra jokingly introduces himself to Leslie Crenshaw as a "doctor specializing in reproductive endocrinology." Leslie begins to divulge confidential health information and is harmed when John explains that he is really, "just a student."

There are some confusing terms that were sometimes used in the past but are now discarded as confusing, such as "student-doctor" and "student-nurse." Some clinicians might still use misleading titles with the paternalistic assumption that it puts patients at ease. But this is no longer considered an acceptable reason for deception. If you are introduced as a "doctor" or a "dentist" or a "nurse," it is best to have practiced what to say in advance, such as, "Oh, thank you for the compliment, but not yet. I am still in school."

# 1.5 Social Networks

WEBSITES SUCH AS FACEBOOK, FLICKR AND YOUTUBE are pervasive in today's culture. Many people use image/video hosting and social networking websites to communicate with family and friends. However, these communications, including pictures and videos, are often not private and confidential. They are, therefore, available for unintended and in many instances, unwanted audiences. A seemingly innocent comment about a colleague on a personal blog, a Facebook page or an unflattering picture on Flickr can reach millions of people in seconds. Even if you try to take it down or edit the content, the cat is out of the bag ... possibly forever.[1]

In Act 1 of *The Brewsters*, Cheryl and John experience some of these unintended consequences by allowing inappropriate photographs of themselves from a pool party to be posted on Flickr. They fall into the hands of parents who report it to the Dean.

Students (and practitioners) need to be mindful of how they depict themselves on social networking websites, blogs and image/video hosting services. The reputations of individuals and institutions are at stake. What is more important, though, is that students and health professionals interact and have access to patients and research participants and their records. Their information may not be shared regardless of medium.

Because social networking media are of recent origin, few institutions have official policies about their use. You might, therefore, think that there are no rules or guidelines concerning their professional use. Not true. Social media have enough in common with other activities, such as sending emails or publishing opinions in blogs and journals, that it is clear what a healthcare professional can and cannot do on social networking websites. Foremost, you must always preserve the privacy of any patient or research participant. You should never post a person's photo taken in a private setting without their permission. You also may not publish any identifiable patient photo without a signed patient release. Posting a recognizable (by themselves or by others) photo of a patient or research participant without their permission is a breach of confidentiality, and a betrayal of the trust they have put in you.

What to post about yourself is a bit trickier. It is more a matter of good judgment and maturity. The best advice is, don't post anything that you wouldn't want your patients, professors, Deans, or future employers (including internship and residency directors) to see — today and thirty years down the road.

# 1.6: Academic Integrity & Academic Misconduct

ACADEMIC INTEGRITY IS THE FOUNDATION of university life. It begins with integrity itself, which is a quality of character, a consistency of action with regard to beliefs, values, and principles. Academic integrity means acting honestly and with respect for teachers and colleagues. It means respecting knowledge, truth, learning, studying, research, scholarship, and writing.

Academic misconduct or cheating is any behavior which violates these principles and the policies that enforce them. The major forms of academic misconduct are plagiarism, the fabrication or falsification of data, cheating or colluding with those who cheat on exams, and professorial misconduct. There are many reasons why cheating is not allowed. Cheating undermines an accurate evaluation of your true abilities, potentially endangers future patients and research participants, and calls into question the validity of published research.

If someone in a lab falsifies data, fabricates data, or omits negative results, then every researcher who follows will have been led down a false path. Scientists are expected to make sure their results are repeatable by other labs — this is part of the definition of scientific methodology. If conclusions are the result of any sort of cheating, they undermine the very fabric of science.

Cheating in research may involve fraudulent use of federal funds, i.e. the research grant that funds the lab. Discovery of such problems can lead to consequences for the entire university, especially if an investigation leads to the conclusion that the university did not do enough to prevent the fraud. In such cases, the careers of others working on the same project can be jeopardized.

Plagiarism is defined as copying or using the words or ideas of another without acknowledging their source. At colleges and universities, plagiarism — whether committed by students, professors, or researchers — is understood as a form of academic fraud or dishonesty and can be punished in various forms, up to and including expulsion. According to U.S. law, original works are automatically entitled to copyright protection the moment they are written down, and violation of copyright can result in legal action.

According to Plagiarism.org, an anti-plagiarism website, all of the following are considered forms of plagiarism:[2]

- Turning in someone else's work as your own

- Copying words or ideas from someone else without giving credit

- Failing to put more than three words in a row copied from someone's work in quotation marks

- Giving incorrect information about the source of a quotation

- Changing words but copying the sentence structure of a source without giving credit

- Copying so many words or ideas from a source that it makes up the majority of your work, whether you give credit or not (see our section on "fair use" rules)

Plagiarism is perceived by many faculty members as a growing problem due to the Internet. The concern is that students today are accustomed to very easy "research." A student can "Google" something in seconds that used to take a walk to the library, digging through a card catalog, finding books, and browsing through them to find a useful passage, or taking notes by hand. It is widely feared that, as a result, students today don't fully appreciate the work that went into the writing they read online, and the ease of cutting and pasting makes plagiarism seem less serious. However, there is no doubt that the ethical violation is the same, even if it is easier to do. "Don't cheat" includes "don't plagiarize." Honesty requires always giving others credit who helped produce a work and providing references for anything quoted or paraphrased. That is just as true whether the words used are copyrighted or in the public domain.[3] To combat the growing use of web-based plagiarism, many schools now use software to automatically detect this form of cheating.

Cheating on exams is any form of deception, dishonesty, or violation of school policy that interferes with the evaluation of knowledge or skills. Collusion also is a form of cheating and includes having someone else take your test, write your paper, or provide you with help when the assignment was meant as an individual task, not group work. Collusion includes both providing help as well as receiving help. It is important to ask your professor what is allowed and what is not, since group assignments ("team-based learning") and group study or note sharing are sometimes encouraged to appreciate the importance of working in a cooperative manner as a team member. Nevertheless, copying an exam is never allowed. If a professor wants you to use a previous exam to study, the professor will announce it and tell you how to access it. But if not, it is safest to assume that looking at an earlier exam is cheating. It is also cheating if a group of students memorizes questions and answers and meets to reassemble the exam after taking it.

Professorial or faculty misconduct encompass all forms of academic cheating that apply to students. It also includes any improper evaluation of students, intellectual dishonesty in the classroom, taking credit for students' research or writing, as well as sexual harassment or other forms of abuse of students. All schools have policies dealing with professorial or faculty misconduct. Sexual harassment and grievance policies provide guidance for student reporting.

Each school has clear and detailed policies concerning student cheating. Some include an Honor Code as well, in which students either recite or sign a pledge before school begins and/or before each exam. Anyone who is aware of another student cheating has an obligation to report it to the appropriate person in their school, such as a faculty member or administrator.

# 1.7: Self Regulation

SELF REGULATION IS A RESPONSIBILITY OF PROFESSIONS as a whole and of the individuals privileged to enter them. Society has delegated to professions the legal authority to define themselves and to determine which individuals are qualified to enter and remain members. Each profession constructs its own licensure exams, which require mastery of that discipline's knowledge and practices. It also creates requirements for continuing education, regulations, and standards which, if violated, result in some form of sanction, such as rehabilitation and the possibility of expulsion. Members have a responsibility to report individuals whose behavior violates these standards. These standards include ethical behavior based on altruism and the fiduciary responsibility to put patients, science, and the public interest ahead of personal self interest.

As an individual, you have a personal responsibility for self regulation, which includes self care. That is, you have a duty to avoid behavior that puts patients or colleagues or the public interest at risk. Unethical behavior may include excessive or binge drinking, the use of illicit drugs, or any activity which might impair your ability to work according to the best practice standards of your discipline.

The altruistic obligations of professionals carry with them an obligation of self care. If you cannot adequately care for yourself, you will not have "a self" who is adequate to serve others. Burnout, impairment, and depression are not uncommon experiences in professional education and practice. They can lead to poor performance, scientific or clinical mistakes, and even injury to patients or colleagues. Maintaining a work-life balance is one of the most important forms of self care. Self care includes: attending to your close relationships; monitoring stress levels, and using stress-reduction techniques; getting adequate nutrition and exercise; maintaining other interests; and continuing your religious or spiritual life. It also is important to ask for confidential help from counselors, mentors, and other members of the health professions.

Self regulation and self care require maintaining an enjoyable private life as well as an altruistic, self controlled professional life. You are not required to live a self-denying monastic life.

In Act 1 of *The Brewsters*, first-year medical student Walter Brewster drinks more than a moderate amount of beer. The resulting loss of self control leads to an automobile accident and injury to himself. His fellow students were faced with the question of whether to take his keys or allow him to drive. In a clinical or other setting where his behavior put other people at risk, Walter's behavior could have had even more serious consequences.

# 1.8: Duty to Report

PERHAPS THE MOST DIFFICULT PART of professional codes involve the duty to report other members of the profession or fellow students. This responsibility is present in codes of ethics for good reason. To be granted the right to be self-regulating, a profession makes a contract with society to set high standards and enforce them. The American Nurses Association, the American Dental Association, and the American Medical Association, for example, expressly state that their members have an ethical responsibility to report incompetent, impaired, and/or unethical colleagues.[4,5,6] Investigations are carried out impartially according to specific policies and procedures that are meant to protect members against unwarranted allegations, to protect whistle blowers from retaliation, and to provide forms of rehabilitation when appropriate.

If a doctor, dentist, or nurse has a drug or alcohol problem, his or her patients can be endangered. It would be unethical if not criminal not to report this until a patient is significantly harmed or dies. To encourage immediate reporting, most professional groups have a system where an addicted, impaired clinician who is reported will not lose his or her license so long as counseling is agreed to. The clinician will typically go through counseling and enter a rehabilitation program with regular testing to get a clean bill of health. There also may be a followup, such as meeting with a professional mentor once every month for a year and then every six months the following year.

Sadly, every year there are people who don't get reported early enough, are in denial, or fear embarrassment or punishment. These may be the cases most likely to lead to either significant harm or death of a patient, or sometimes the death of the clinician (an accident, overdose, or suicide). Hence while the duty to report serves its primary purpose of protecting patients, it also might prevent harm or expulsion. As hard as it is to report a fellow professional, you are protecting patients, upholding the integrity and reputation of the profession, and perhaps doing the person a favor as well.

These responsibilities apply to students as well as to established professionals, and for many of the same reasons. Cheating on an exam, for example, undermines evaluation of your true knowledge and competence, potentially endangering future patients and research participants. In a school of biomedical science or public health, cheating calls into question the validity of published research (and the careers of others working on the same project).

In some schools students take an Honor Pledge or an Ethical Pledge, which makes it official policy to report academic misconduct. For example, a student who has taken an Honor Pledge and observes or suspects cheating on an exam is required to report this privately to a course director or department chair, who will report it to the Dean of Students. Depending on the school, the student may not be expelled, but will be disciplined and warned about the importance of improved behavior in the future. Students who witness sexual harassment, or forms of unethical behavior which put others at risk, also are strongly encouraged to report it to a professor, department chair, dean, or the Office of Institutional Compliance. In any case, even in schools without an Honor Pledge, it is the duty of healthcare profession students to report misbehavior of their peers, just as it will be their duty to report their peers when they graduate and enter practice.

# 1.9: Confidentiality

CONFIDENTIALITY IS OFTEN THE FIRST KEY CONCEPT learned by students in clinically based disciplines. It's an early chapter in most textbooks for two reasons: before you learn anything about a patient, you must understand why you have a right to the patient's personal information and why you have a duty to maintain confidentiality. Personal information can be defined this way: if you would not have known the information for reasons other than being a health professional (or health professions student), then you cannot share that information with anyone except those involved in the patient's clinical care. Confidentiality is also an important ethical idea in clinical research, public health, and bioinformatics.

Beginning students may not appreciate at first the importance of confidentiality. Like honor codes, the duty to report academic misconduct, informed consent — even dress codes — confidentiality is an element of professionalism that must be learned and followed. When you become a health professional, you are changing your identity. Your new role requires ethical duties and competencies that will change your understanding of yourself as well as the understanding others have of you.

A physician, nurse, or dentist may not disclose any medical or personal information revealed by a patient undergoing treatment. The requirement of confidentiality has a long history in medicine. The responsibility is based on the idea that a patient must feel free to disclose information and speak openly to be properly diagnosed and treated. In return for this honesty, the clinician must not disclose confidential information or communication without the patient's expressed written consent, except when required to do so by law. The trust engendered by confidentiality is itself a means of healing in the clinician-patient relationship. Maintaining confidentiality is also a requirement for anyone conducting or participating in experimental research with human subjects.

Confidentiality is more difficult to maintain in an age of third-party payers and electronic medical records. Patients routinely sign releases allowing information to be disclosed to family, friends, various healthcare providers, and to insurance companies. But insurance companies often place patient information in a database that may be accessed by employers or other insurance companies without consent of the patient. Electronic medical records, which improve collection and sharing of medical information, also may lead to serious breaches of confidentiality.

With the push to digitize patient health data, the federal government has enacted new and more aggressive regulations to protect patient health information. While these regulations help protect confidentiality, they cannot ensure complete compliance as long as human beings are involved in the practice of health care. Without the ethical behavior of nurses, doctors, residents, students, and other healthcare workers, confidentiality can be breached from any computer terminal.

There are legitimate exceptions to the duty to maintain confidentiality, which may require notification to law officers or public health officials. These include injuries caused by weapons or crimes, child abuse, elder abuse, domestic violence, and highly infectious disease or partner notification to public health officials. Despite these complications and exceptions,

students must always assume that it is their duty to protect
confidentiality.

# 1.10: Sexual Relations & Harassment

WHEN STUDENTS ENTER THE HEALTH PROFESSIONS, their fiduciary responsibilities take precedence over all else — including sexual feelings, interactions, and relationships. It may not be immediately obvious why this is so.

Normally, sexual relations between consenting adults are a matter of personal freedom. However, things change when your job is to promote the health and well-being of others. When a clinician is responsible for the care of other people, there can be no expression — not even a hint — of sexual interest whether by touch, word, innuendo, facial expression, or body language. Patients allow caregivers physical intimacy that is not permitted to anyone except spouses or significant others. When they tell secrets, disclose personal information and express intimate feelings, they trust that clinicians will keep those conversations confidential.

If a patient or a research participant expresses any sexual interest in a student or clinician, it should be ignored. If it continues, it should be turned aside or labeled as inappropriate. One's focus should remain on the work of the case. If any form of sexual interest continues, the student steps aside and lets others care for the patient or, if necessary, reports it as sexual harassment.

Sometimes a student will have troubling sexual feelings or fantasies about a patient. This does not mean the student is a bad person or is unprofessional. The question is: what to do with the feelings or fantasies? First, there can be no acting on them. When state medical boards review unprofessional behavior, sexual misconduct is one of the common problems and this may lead to permanent loss of the license to practice. Second, being aware of and acknowledging sexual feelings or fantasies can be an important part of growth. Increased self-understanding and relief from guilt can come from discussing them privately and confidentially with other students or faculty or loved ones. If sexual feelings or fantasies about patients persist, one should seek counseling.

Just as a patient is in a vulnerable position with a clinician, a student is in a vulnerable position with a faculty member. In both cases, the imbalance of power increases the potential for exploitation and harm. Sexual relations between students or clinicians and patients are strictly prohibited. Sexual relations between faculty and students are not explicitly prohibited, but they are strongly discouraged, given the potential for abuse or exploitation. Moreover, any unwanted sexual advance, comment, or innuendo could be a form of sexual harassment.

Sexual harassment in any form is unethical and impermissible. Students who feel as though they have been sexually harassed are encouraged to contact the appropriate dean of their school for additional guidance and assistance.

# 1.11: Accepting Gifts from Industry

THE QUESTION OF WHETHER TO ACCEPT GIFTS from pharmaceutical and medical device companies has been the subject of numerous articles and books during the past 10 years. What is considered acceptable has changed dramatically, as research demonstrated emphatically that gifts create a sense of gratitude from the person who accepts them. The scale of pharmaceutical company marketing is staggering, with estimates between $20 billion and $60 billion per year,[7] 50% of which is free samples and 25% "detailing" physicians.[8] The industry employs thousands of sales representatives who provide gifts to physicians, and more recently to nurse practitioners authorized to prescribe drugs.

As part of their marketing efforts, pharmaceutical companies provide meals or hospitality. They also sponsor medical education and educational meetings for physicians. They can give gifts bearing company names, ranging from pens, writing pads, or flashlights to cameras, watches, medical books, and equipment such as reflex hammers and instruments, such as the fictional "Hittman" stethoscope Cheryl has the choice of accepting or rejecting in Act 1 of *The Brewsters*.

A common belief of those who accept gifts is that one can remain objective. Yet studies have shown that a physician's prescription of a drug being promoted goes up after receiving gifts from the manufacturer.[9] Research also indicates that medical students often do not appreciate the influence of drug company gifts and advertising. The value of the gift may not be important, a $5 pocket flashlight to test papillary reflexes can be just as effective as a $300 stethoscope.

In 2002, the American Medical Student Association launched a "PharmFree Campaign" to reduce conflicts-of-interest at medical schools and academic medical centers. [10] Its leaders urged medical students and all healthcare providers to seek out unbiased and evidence-based information, rather than relying on pharmaceutical company representatives and publications. A few years later, a national survey of students' exposure to and attitudes about drug company interactions found students at risk for "unrecognized influence by marketing efforts." [11] Medical schools and the American Association of Medical Colleges began tightening or developing policies regarding gifts, meals, access to campus, samples for patients and for physicians, and disclosure of financial *conflict of interest* (COI).

Accepting gifts and other benefits from industry raises important issues. In today's revitalized climate of professionalism, arguments against accepting gifts are compelling: accepting gifts impairs objectivity, creates the expectation of reciprocity, increases the cost of health care, weakens fiduciary responsibility, and generates the perception of a conflict of interests.

# 1.12: Cultural Understanding

ALL HUMAN BEINGS TEND TO THINK they and their family, religion, and cultural traditions are normal. That is, we think our own morality alone is true and right. This can easily lead to an inherent bias against others who come from a different background. The United States is a pluralist society in which no religion, race, or ethnic group is privileged above others. This pluralism extends to health care. In a pluralist society, health care is delivered without assuming the values of any particular culture. Fiduciary responsibility requires humility about your own culture and respect for the values and choices of patients making decisions in accord with their core beliefs.

Of course, Western medicine can be seen as having its own culture, which traditionally focuses on biomedical aspects of disease to the neglect of psychosocial and cultural factors in health. Today's emphasis on cultural competence, or cultural humility, attempts to redress this imbalance. Cultural competence is one element of professional competence. It is the ability to function effectively in providing care to patients with diverse beliefs, behaviors, and needs. The goal is improving health care by establishing relationships based on respect for and understanding of cultural differences.

It would be helpful if all healthcare professionals were familiar with the world's great religions: the Abrahamic religions of Judaism, Christianity, and Islam; the religions of East and South Asia, and the religions and world views of indigenous people in Africa and North and South America.

Culture, though, is more than religion. The term generally refers to the customs, beliefs, values, languages, and institutions of any racial, ethnic, religious, or social group. It is, of course, impossible to be fluent is all cultures and religions, not to mention the languages which express them. But familiarity and unfailing respect are essential for improving quality of care for diverse groups.

Cultural competence is also a strategy to reduce health disparities. Culture and identity are embodied in social groups which include not only race, ethnicity, and religion but also age, sexual orientation, gender, disability, and socioeconomic status. Linguistic minorities are another group and include people with low literacy skills, limited English proficiency, and the hearing impaired. Minority social groups are especially exposed to the power imbalance between healthcare professionals and patients. There is a long history of stereotyping, institutionalized racism, and other forms of devaluation and discrimination. Cultural competence in clinicians and in culturally sensitive institutions can help minimize these imbalances.

# 1.13: Conscientious Objections

A CORE ASPECT OF PROFESSIONALISM is that you act in the patient's best interest, putting personal opinions aside to care for patients in fulfilling your professional responsibility. However, there are times when professional responsibility conflicts with personal morality. Performing a certain action could violate your sense of integrity or conscience.

Conscience is an inner voice or gut feeling that something is right or wrong. The demands of conscience and personal morality are rooted in faith or tradition rather than reason. Personal *morality* contrasts with *ethics*, the philosophical field of study in which actions are evaluated by rational argument alone.

*Conscientious objection* in healthcare is the notion that a provider may legitimately act against patient wishes or medically indicated treatment because of personal moral values. If asked to perform an abortion, for example, a healthcare professional might refuse based on her conscience.

It is widely accepted that conscientious objections to terminations of pregnancy and sterilization should be honored. But to what extent should personal morality be allowed to override professional obligation? What about a pharmacist who refuses to dispense contraception or a "morning after" pill? Where do we draw the line? Should a trauma surgeon be allowed to refuse treating the injuries of a reckless drunk driver?

There are deep divisions about the permissibility of conscientious objections. Some argue that conscience has no place in the delivery of health care and that professionals should follow ethical norms rather than personal morality. Others argue that if a conscientious objection compromises quality or efficiency or equitable (just) delivery of health care, it should not be permitted. Some say that providers who cannot or will not offer legally permissible and potentially beneficial care should not work in healthcare or should choose a field that does not pose such problems (for example, do not go into Obstetrics or Women's Health if you feel you could not help a woman who requests abortion, contraception, or sterilization). Most professional societies — the current consensus — would permit only a tightly circumscribed range of objections to performing certain actions, but still require professionals to provide full information and referral for those services.[12]

It can be said that the fiduciary responsibility of healthcare professionals — their duty to put patient welfare above their own interests — can sometimes be in tension with personal morality. However, since the purpose of the health professions is to promote health, cure disease, and relieve suffering, personal morality generally must be subordinated to professional responsibility.

TURN TO PAGE 163 TO READ ACT 2.

# ACT 2

"No man is an island."

*~ John Donne, English clergyman & poet, Meditation XVII*

*(1572 - 1631)*

# Accident

THERE IS A RUSH FOR THE RUINED CAR. You have a head start, though, and arrive first. There is the odd reek of spilled antifreeze, but no gasoline smell, which makes you feel better. You trot around the vehicle to the driver's side to be greeted by a surprising sight. Walter is trying to crawl out through the broken window. You see some cuts and some evidence of blood, but no limbs at a crazy angle. You pry open his door. He falls out of the car and rolls onto the pavement. Walter begins to unsteadily stand as you reach his side.

"Take it easy, Walter," you tell him. "You should probably lie down over here on the grass until we can check you over. Does your back hurt? Any neck pain?"

He looks at you, blinking in incomprehension. "What? No. Why?" Then his head swivels to take in the wreck. "Oh, man. Look at my *carrrrrrrrrrrrrr!*"

There is a bit of commotion and you glance at the growing crowd. Pushing his way through the sea of camera phones is Dr. Hernandez. He strides to Walter's side and takes his arm, gently pulling him to the side of the road. "Come on, Walter," he says. "Come and sit down for me."

Walter allows himself to be guided, but is still staring at the wreck. "My car ..." he says to no one in particular.

"Your car's a little outside my specialty," Hernandez tells him. "You are not. Please sit down."

Walter sits heavily in the grass. You watch as Hernandez carefully examines Walter as best he can in the fading light, cautiously moving his head, his arms and legs. Walter continues to stare at his smashed auto. Julie Spates sits down on Walter's other side and takes his pulse.

"I can't believe this," Julie mutters. "Hey Cheryl," she calls to the crowd. "Anything here remind you of your days in the children's emergency room — especially the *children* part?"

Hernandez looks up, grinning.

Julie continues. "Shoot, maybe we should take him there. It's not far."

Oddly, you realize you haven't been conscious of the ambulance's siren until it's half a block away. You see some of the students on the outskirts of the crowd waving to the flashing lights, then people begin to part, clearing a path.

Walter's beginning to exhibit more awareness of his surroundings. He hears the ambulance and gets agitated. "No ER!" he tells the doctor. "I don't wanna go to the ER!"

"I'm going to agree with you," answers Hernandez. "No need for the emergency room. I think you're okay. I can patch you up at my clinic and let these guys attend to someone who's actually hurt." The doctor stands up from his examination of Walter and turns to you. "I'll go talk to the EMTs, let them know what's going on. Keep an eye on him, will you? Yell if anything happens, though I doubt it will."

You watch as Hernandez walks toward the ambulance, waving. It looks like at least one of the EMTs knows him. You turn toward Walter, still seated in the grass next to Julie, but somehow looking smaller and smaller, as if he were collapsing in on himself, his eyes fixed on the car wrapped around the tree. The buzz of conversation around you grows in volume. One phrase in particular seems to leap out of the background noise: "You know, I heard it's almost impossible to kill a drunk driver …"

Finally, Walter's gaze turns to you, his eyes confused and lost. "How'm I going to get home?" he asks.

Fortunately, Dr. Hernandez is back at your side, so you don't have to answer. "You're not going home yet. I'm driving you to my clinic first so I can do a better job of checking you out." He turns to Julie. "I can use an assistant." Julie nods. Hernandez holds out his hand to Walter. "You can walk, right?"

Walter takes the offered hand. When he seems to be having a bit of trouble, you take his other arm and steady him. Julie pushes from behind. Leaning on Hernandez, Walter takes a few steps, limping slightly. He pauses a moment, looks at you and says, "Thanks." It's a surprising moment for you; Walter seems totally sincere, almost sober.

Julie says, "I'll run and get my bag." Walter and Hernandez move slowly through the crowd.

You follow them as they limp toward the house — with each step, Walter seems to be getting a little steadier. Carla meets them just outside the house, gesturing with her cell phone. "I called Sheila and Wayne," she tells them. "Wayne will meet you there."

"Great," Walter mumbles. "Like I needed that, too."

Hernandez helps Walter into the passenger side of his SUV. "Walter, you're a very lucky young man tonight," he says. "Don't try to pretend it's otherwise." With that, Hernandez climbs into the SUV and you watch him pull out and drive into the night. Julie is a car length behind. The day suddenly seems exceptionally long and you reflect that it will be good to be in your bed. But, as you look at Walter's wrecked car down the street, you also consider that there's no particular need to hurry home.

TURN TO PAGE 169 TO CONTINUE.

# Answer the Phone

ENRIQUE HERNANDEZ, MD turns his SUV into the parking lot in front of his clinic and notes a Jeep waiting there. It's Walter's father, Wayne Brewster. Hernandez is not surprised that Wayne got there first. Besides the speed a worried father would muster, there was also the matter of the two times Hernandez had to pull over so Walter could vomit. There was no blood evident in the expulsion, which was good, at least.

Hernandez motions to the Jeep to follow him. The Jeep and Julie's Toyota continue around the building to the clinic's back door. While Hernandez unlocks the door and strides within to turn on the lights and turn off the burglar alarm, Julie guides Walter into the clinic. Wayne, a man settling into late middle-age with a thickening waistline and a thinning hairline, joins them.

Wayne looks at his son. "How are you doing?"

Walter looks like he wishes he could vomit again, on cue. "I'm okay."

Wayne makes a motion to join Julie in supporting Walter on the other side. "You need some help?"

Walter waves him off. "I said I'm okay, Dad," but continues to accept Julie's aid.

Hernandez ushers them into an examination room and helps Walter onto the table.

There is a sudden electronic chirping, and Hernandez, Julie and Wayne all take out their iPhones in unison, a moment that seems almost choreographed. No one has bothered to change their phone's default ring. No one knows whose phone is ringing. The humor isn't lost on the three, who smile at each other in the absurdity of the moment. A bit of tension exits the room.

TURN TO PAGE 207 IF JULIE'S PHONE IS RINGING.
TURN TO PAGE 171 IF WAYNE'S PHONE IS RINGING.

# Wayne: My Phone

OF COURSE, IT'S SHEILA ON THE LINE. "Wayne, are they there yet?"

"Yeah," you reply, "they just got here."

"So how is he?"

You watch Julie, Hernandez' nurse offer a clipboard to Walter with some form or other. Walter slowly takes the pen from her and signs it. It looks to you that Walter and Julie are friends, or at least know one another. Then it dawns on you that Hernandez told you last year that his clinic nurse, Julie, was going back to nursing school for some kind of advanced degree. *This must be her.* You figure Julie and Walter must see each other around campus from time to time.

"He says he's okay, Sheila," you say. "He looks a little shaky and beat up, but otherwise fine."

"Good. I didn't think it was like Carla to sugar coat things."

"No," you say absently, watching Dr. Hernandez shine a light into Walter's eyes, then flick it away, observing his pupils.

"Is he going to have to go into the hospital?" Sheila asks flatly. You can tell she's distracted by whatever's on the TV you can hear in the background. Chances are she's also taking advantage of your absence to smoke in the house.

"Well, the doctor hasn't said anything yet, but unless he finds something wrong, I don't guess he will."

"That's good. Call me if there's anything like that."

"Sure. Cindy in bed yet?"

"I think so. She was worried about Walter and didn't want to go, at least that was her excuse." You hear Sheila shout across the house to your older daughter. "STEPHANIE! IS CINDY IN BED YET?"

Well, if she was, *you just woke her up*, you think. Dr. Hernandez and Julie are now cleaning out the cuts evident on Walter and bandaging the worst of them.

Sheila has returned to the phone. "Yeah, she's in bed. I had Stephanie tell her Walter is okay. He *is*, right?"

"Yeah, he seems to be. I'll call you if need be."

"Okay." Click. *Yeah, well, don't say goodbye or anything.*

Hernandez is checking a hypodermic Julie has prepared. "Tetanus booster," he says, apparently to the needle, but you know it's to you as much as to Walter. You look away as the needle slips into Walter's arm. Hernandez continues, "While we're sticking you, I want to take some blood."

Walter waves him off. "You don't need to do that."

"Walter, I think you're okay, but a white count would help me know if there's any internal damage that's not obvious."

"Yeah, but ... nah, I'm okay," Walter says as he struggles to a sitting position.

Hernandez considers him. "Look, you know HIPAA prevents me from giving patient information to the police. That includes blood alcohol levels."

"I'm good," Walter insists.

"Right," Hernandez turns slightly, so he can speak to both you and your son. "I can't find any evidence of a concussion, but it's not unheard of for symptoms to show up two or three days after a trauma. We're going to have to keep a close eye on you. You were very, very lucky tonight, Walter — but you're not going to feel very lucky for the next few days. I'm not going to lie to you, you're going to be sore as hell. How are you set for Ibuprofen?"

"Got plenty of it."

"Good. Take as directed. And, either of you let me know if there's any change, right?" You both nod. "Okay then. He's your problem now, Wayne."

You nod again and say, "Yeah," a bit too ruefully. "Let's get you home." You help Walter down from the examination table, and move him slowly toward the back door. You hear Hernandez' voice behind you, "Hopefully I won't see either of you for a while."

You turn back with half a smile on your face. "Sorry, Doc, I'll see you later this week. I have an appointment on Tuesday."

Hernandez seems almost amused, but not at all surprised. "Really? Okay. I'll see you then."

You and Walter hobble into the night air. He seems to be looking forward to this drive home just as much as you are. Which isn't much.

TURN TO PAGE 174 TO CONTINUE.

# Wayne: The Drive Home

SILENCE IS A FUNNY THING. You crave it at home, in your off-hours, but you never get it. Now that you have it — in the car, between you and Walter, as you drive to your house — it's the worst thing in the world, a smothering blanket on a hot summer day.

It's Walter who finally breaks the silence. "I want to go to my apartment."

You shake your head. "Dr. Hernandez says we need to keep an eye on you for a while. Besides, you've never been in a wreck before — I have. Moving around is going to be a real chore for a few days."

Walter sighs. "I'm not lookin' forward to what Mom has to say."

You feel a pang of sympathy for the boy. Well, not a boy anymore, at least in theory. He's your firstborn, your only son. Sheila will likely hug him hard enough to produce a groan from his battered body, and then there is going to be a firestorm of never-ending complaints that's going to make the kid wish he had died in that wreck.

"Your mother might surprise you."

"I doubt it."

The silence threatens to come back. Trying to hold it at bay, you ask, "So why?"

"Why what?"

"Why are you drinking so much?"

"I'm not —"

"Come on, Walter. You ran your car into a tree, and you smell like six packs and puke."

Walter lets out all his breath in an explosive rush. He seems to be considering what to say, but finally says only, "You wouldn't understand."

"Try me."

"Mom and Nani always think I'm going to save the world, you know? I'm the Golden Child, I can do no wrong." He snorts miserably. "I've done plenty wrong."

"You've done a lot right, too. You got into medical school, that's not nothing."

"No, it …" Walter gestures uselessly with his hands. "Now it's like everything's doubled, you know? The pressure, the expectations, the chances to screw up …"

"And the chance to succeed," you say, hoping you sound wise.

"Oh, *thank you!*" Walter explodes, "That is exactly the sort of meaningless motivational poster crap I need to hear right now!"

The silence comes back in a hurry. You could use a beer yourself right about now, but you reflect that would be a bad message.

"Look," you say, finally. "You've worked hard to get here. Your mother and I have worked hard to get you here too. Your Nani is more proud of you than you can possibly know. I don't want you throwing all that away like this."

You're at a stop light now. You look at Walter staring out of the window, either deep in thought or trying to keep from throwing up again.

"Walter," you tell him, "I just don't want you winding up like me."

Walter continues to look out the window. "Don't worry, Pop," he says. "That's the last thing I want, too."

*That hurt.* The light changes, and you drive on. The silence is back, but at least it seems to be a little lighter. At least, you hope it is.

TURN TO PAGE 177 TO CONTINUE.

# Wayne: DEXA scan

YOU HATE DOCTOR'S OFFICES, even when the doctor's a good one, like Dr. Hernandez. So why do you seem to be in them so often? *Well, Wayne,* you consider, *you hate hospitals even more.* You're here today to make sure you don't wind up in some hospital.

You look around the examination room. Hernandez has six of them, and you've been in them all. This is number three, the one with the abstract art print in shades of brown. You like this room the most. Not at all like number six with its creepy, brightly colored clown photograph. You know that's the room for young patients, but man ... you hate clowns.

There is a soft rap at the door and Dr. Hernandez and Julie enter. "How's Walter doing?" the doctor asks.

"Doing well, thank you for asking," you reply. "Surprisingly well. He's moving around more, and he's talking about going back to classes. Though I guess that's more due to his mother driving him nuts than anything."

Hernandez chuckles and pulls his rolling stool from under the small desk along the wall. He sits and regards you. "So Wayne ... what's the story?"

"Okay," you begin, "it's like this. A few weeks back, my mother-in-law, Barbara Ann? Anyway, she had one of those DEXA scan things."

Hernandez nods. "For osteoporosis."

"Right," you confirm. "Anyway, it turns out she *does* have osteoporosis, and now she's taking … well, I'm not even going to try to tell you the medicine's name without the bottle in front of me."

"Yes," Hernandez says, "we're treating it." He pauses, waiting for you to continue.

"Okay," you say finally. "I just turned 50 not too long ago, you know? I'm starting to notice these aches and pains more and more, particularly after I move some of those big packages and containers of mail. Sheila's always riding me about my posture, says it looks like I've been defeated, or that I look like a shrimp. I'm starting to worry that I might have osteoporosis, too, and I thought maybe I should have a DEXA scan … too."

"You think you have osteoporosis?"

"Maybe just the beginnings of it? I mean, if I do, I should probably start taking the medication now, right? Before it gets too bad?"

Hernandez scribbles something in a folder. When he's finished, he looks up. "Let me check on a couple of things, Wayne. I'll be right back."

And with that, they're gone, leaving you alone with the brown abstract print. You reflect you might like exam room four better, the one with that watercolor painting of the woodpecker. You always liked watercolors.

There's a tap on the door and Hernandez reenters. "Okay, Wayne," he says. "I checked, and we have some time open on the DEXA scan this afternoon. However — there are some things you should consider."

"What? It doesn't hurt or anything, does it?"

"No, nothing like that. I wouldn't ask you to do anything like that, especially if I thought it was unnecessary. That's the main point, Wayne: some insurance companies will pay for the test if you have any of the risk factors. Frankly, you don't. If you insist, I'll give you the test, if only for your peace of mind. But you need to know that you may be out-of-pocket for it."

That makes you stop. You're not exactly rolling in money at the moment, with Walter in medical school, and Stephanie rapidly approaching the age where you have to think about college for her, too. Then, there's the nagging doubt that's been preying on you for some time now. Are your bones turning traitorous or not?

TURN TO PAGE 180 TO ASK FOR A DEXA SCAN.

TURN TO PAGE 183 TO FORGET THE DEXA SCAN.

# Wayne: The Scan

FINALLY YOU SAY, "You know, I've got plenty of other things to lose sleep about, and I think this will at least take care of one of them. I still want to do the test."

"So," Hernandez says. "Are you currently taking any calcium supplements?"

"No. Should I?"

Hernandez shakes his head. "Couldn't hurt, but you're not supposed to take any 48 hours before a DEXA scan. Okay, come back at 3:00 PM and we'll do the deed."

You already feel quite a bit of relief.

The scan that afternoon goes quickly. You're almost disappointed — you expected a bit more of a production. You guess it says something about the state of the technology that it all went so quickly and smoothly, *right?*

The next day, you receive a phone call from Hernandez. You look at the display of the iPhone, gut churning, thinking. *This is stupid, Wayne. Answer the phone, then it will all be over.* Finally, the chirping of the phone stops, and after a moment, the voice message icon appears. You thumb the button and wait for the voice mail to play.

It's a pleasant, female voice. "Hi, Wayne, this is Julie at Dr. Hernandez' office. We got the results of the DEXA scan and I'm pleased to tell you that you definitely do not have osteoporosis. Have a nice day!"

You take a deep breath and make a silent prayer of thanks. That's good. That's definitely good.

Wait a minute — *then what's with all the aches and pains?*

TURN TO PAGE 182 TO CONTINUE.

# Wayne: Insurance Won't Pay

THREE WEEKS LATER, your shift has just ended and you sit in your Jeep, feeling the ache in your bones after a long day, wondering if you're in the mood for talk radio on the drive home. Maybe some classic country. Something from your childhood, with its easy familiarity, a reminder of a time you didn't have to worry so damned much.

You take your phone from its belt holster and notice the missed call icon. Tapping over the recent calls, you see it's from Dr. Hernandez' office. That can't be good. Dialing into your voice mail, you find that is indeed *not* good.

It's Dr. Hernandez' clinic manger, Lily. She sounds pleasant but a little bored. "Mr. Brewster, this is Lily at Dr. Hernandez' office. I'm afraid your insurance has declined your claim on the DEXA scan procedure and we're going to be sending you a bill for it. The charge for a DEXA scan is $350. We didn't want this to be a surprise. If you have any questions, give me a call at …"

You're not listening anymore. Three hundred and fifty dollars. Sheila is going to hit the roof over this.

You let out a heavy sigh and glance at the clock. 5:30 PM. It's still happy hour. Maybe a couple of drinks at Fatso's Bar will soothe the aches and pains, physical and otherwise.

TURN TO PAGE 177 TO TRY AGAIN.

# Wayne: No DEXA Scan

YOU LOOK AT HERNANDEZ. "So I don't have any of these risk factors?"

Hernandez shakes his head. "None of the classic ones." He starts counting with his fingers. "You're not a woman. You're not post-menopausal. You're not 65 years old. You have no family history of osteoporosis. You're not taking any medications that could lead to osteoporosis." Hernandez drops his hand. "Unless I'm an incredibly unobservant doctor, which I don't think I am, you're not a likely candidate for osteoporosis."

You finally have to laugh, "Okay, okay, I'll take your word for it. But these aches and pains I've been having …"

Hernandez shrugs. "Walter, you yourself said it. You're 50 years old. You don't get enough exercise. You're a mail carrier, you drive around in that little Jeep all day."

"I have to get out and deliver packages to doors," you say, a bit defensive. "So … more exercise would help?"

"Wouldn't hurt." He begins to scribble in your folder again. "Tell you what. I'll put your concerns in your record, and the next time we have you in for a checkup we'll pay special attention to that, make sure you haven't developed any of the warning signs for osteoporosis. Okay?"

"You're the doctor."

"Thank you for remembering." He closes the folder and looks directly at you. "You're doing okay, Wayne. Get some calcium supplements if you're worried, but honestly, this isn't something you should be concerned about. We'll see you next time."

With that, Dr. Hernandez is gone, knocking on the door of the next exam room. You feel a little better, but still have some lingering doubt. Particularly when that familiar twinge hits your lower back when you step down from the examination table.

TURN TO PAGE 185 TO CONTINUE.

# Wayne: Cindy is Sick

IF THERE'S ANYTHING WORSE than having a sick child, you have no idea what it might be, and you have no desire to find out. You sit at Cindy's bedside looking down at your daughter sleeping peacefully, if a bit fitfully.

*My youngest,* you think. *At first I wasn't sure I could even think of her as mine.* You and Sheila had gone through many channels and jumped through increasingly erratic hoops to adopt Cindy from that Korean orphanage. That was nearly ten years ago, and there is no doubt in your mind now that she is your daughter, just as much as Stephanie.

Sheila had seemed enthusiastic about the whole process back then. If not for her, you probably would have given up on the adoption, but she was having none of that. For the first year or so, too, she was carrying Cindy around proudly, enjoying all her friends cooing and playing with the new baby.

The last few years, though, that seems to have cooled off. Sheila seems to have cooled off, too, as she gets increasingly consumed with her real estate job. That's where she is now, out of town, as usual. Leaving you with a sick little girl who's too young to be left at home alone. You brush Cindy's thick black bangs away from her forehead and gently place your hand there. *I feel for you, little girl. Mommy doesn't seem to care much about me anymore, either.*

Cindy's eye flutter open, two almond-shaped pools of the truest hazel you have ever seen. "Hi," she croaks.

"Hi, baby. How do you feel?"

Coughing, "Okay."

"Cover your mouth when you cough, sweetheart." You notice you sound just like your mother when you were young.

Cindy does sound better, and there's no discernible heat to her forehead. You've been at the house with her for three days now. Stephanie's grades are bad enough. You couldn't ask her to take a day or two off from school to mind Cindy. Walter, after that dreadful crash, has thrown himself into his studies with a passion you envy. Which leaves only you. You wonder a lot how your father dealt with things like this. Tough choices. He seemed to have no problem making them, no matter how much it hurt him.

Then, you reflect, your father didn't work for the U.S. Postal Service. He didn't have to deal with periods like now, when labor negotiations had people twitchier than characters in a spy movie. In times like this, management always seems to like showing how tough they can be, with unexcused work absences — like, admittedly, yours — rating high on the How Can We Stomp Them Into The Ground list.

You smooth Cindy's hair back over her forehead, pondering what to do.

TURN TO PAGE 187 TO SEND CINDY TO SCHOOL AND REPORT TO WORK.
TURN TO PAGE 190 TO STAY HOME AGAIN TO TAKE CARE OF CINDY.

# Wayne: Send Cindy to School

YOU CAN'T HELP BUT THINK this isn't the best decision, but what can you do? Cindy was more than a little grumpy when you told her to get dressed, and that coughing is still worrisome, but you remember one of the guys down at the center talking about how the new bug just seems to travel down to the lungs and lingers on, even after all the other symptoms had passed. Kids are made of rubber bands, right? They heal almost overnight. *Right?*

That doesn't make you feel any better as you drive from the school to work. But what else could you do?

TURN TO PAGE 188 TO CONTINUE.

# Wayne: Cindy Sick at School

YOU HAVEN'T BEEN BACK AT WORK A HALF-HOUR before your phone chirps. It's the school, and Cindy has already thrown up once. She's running a fever, and they want you to come pick her up immediately.

You stare at the phone a moment, its blank face offering you no advice. You feel trapped. You gambled on Cindy making it through the day, and you lost. You're stuck here at work, having made your triumphant return after three days of absence. *Aw, crud.*

Focus, Wayne. *Focus.* You always find the answer when you talk through a problem with another person, right? So, who can you talk to about this?

Well, there's only one person. You thumb Sheila's name on your contact list. It rings a few times before she answers.

"What is it, Wayne? I'm in a meeting."

"It's Cindy. She's sick at school."

"Why are you telling me this? She's been sick, right?"

"Well, yeah, but I'm at work right now and …"

There is an exasperated sigh on the other end of the call. "Good grief, Wayne, just make an excuse and leave. I have to go now."

The line goes dead.

*Just like that, right.* Just pick up and leave. Who does she think you work for, the Tooth Fairy? Sweetness and Light Central?

*Okay.* You holster your phone and propel yourself toward the back door. On the way you pass the supervisor's office and poke your head in for the briefest of moments, saying, "Sorry, Joe, I'm throwing up again. Gotta go. Sorry. Bye," and head out the door.

In the Jeep, on the way to the school, you're fuming. *There has got to be a better way to live than this.*

TURN TO PAGE 192 TO CONTINUE.

# Wayne: Stay Home with Cindy

CINDY IS IN THE BATHROOM NOISILY LOSING HER BREAKFAST. Well, there's no two ways about it — the girl's still too sick to go to school. A phone call to Hernandez' office two days ago confirmed it was the ever-popular "bug that's going around" and there was no treatment for it but rest and fluids.

You phone the office again, preparing to use your best worst miserable voice. You wish there was another way to do this. You wish you were smarter and could come up with another way.

TURN TO PAGE 191 TO CONTINUE.

# Wayne: Boss Phones Home

THAT AFTERNOON, THE PHONE RINGS. According to caller ID, it's your boss. You consider not answering, but that's not going to solve anything. If nothing else, you'd be adding cowardice to the list of all your other failings.

You put on your best miserable voice again. "Hello? Oh, hi, Joe. Yeah, you know how it is. You feel okay and then you try to stand up."

You pause and listen for a bit. "So far I'm keeping down some soup okay. I'm hoping I can be back tomorrow."

Then, there it is, with perfect bad timing: Cindy crying in the next room. She's been having stomach cramps all afternoon and isn't shy about voicing her displeasure.

"What? No, that's … the TV. One of those god-awful talk shows. Look, I gotta go, I was wrong about the soup."

You hang up. You answered the phone, but you still feel like a coward. You get up and walk to the bedroom to comfort Cindy.

*Why can't I be smarter? All I had to do was say, yeah, it's Cindy, and she caught it too.*

*Why aren't I a better liar?*

*Why am I such a loser?*

TURN TO PAGE 192 TO CONTINUE.

# Wayne: Reprimand

YOU SIT IN YOUR JEEP. You've sat there for 15 minutes, doing nothing but thinking and staring at the post office loading dock in front of you.

You had to do it. Your next day back at work, you had to go in and tell your supervisor, Joe, that you weren't sick. That you were staying home with your sick daughter.

It didn't go over well at all; even worse than you expected. You weren't expecting an understanding smile and a, "There, there, that's okay!" But, *good grief.* The boss started with docking a week's pay — which was sort of expected — but then there was his going on and on, the doubting your commitment to the job, the level of trust you can be given, how "This is going on your record," and "You can forget about that raise, let alone a promotion." You have worked at the Post Office for 25 years, yet they threatened you with "a mark on your permanent record" like you're a school kid. You have worked at the Post Office for 25 years, and this is all the consideration they show you. *I have worked at the Post Office for 25 years and I am never going to be anything more than a mail carrier.*

You sigh heavily and start the Jeep's engine. It's time for a stop at Fatso's Bar. You didn't get to do that while Cindy was sick, and you've been so busy trying to play catch-up since you came back, trying to make a good impression …

All for nothing.

The beer is cold and good. You talk with the other regulars. Everybody's got complaints about their jobs, and you try to outdo each other with your Boss From Hell stories. You feel better. A little less alone. Your bones hurt a little less.

*And holy hell, when did it get to be midnight?*

TURN TO PAGE 194 TO CONTINUE.

# Wayne: Mom's Lesion

IT'S EARLY EVENING THREE DAYS LATER, and Stephanie and Sheila have just gotten home. You don't mind cooking the evening meal that much — Cindy just gets irked by your limited range of recipes. She seems to prefer when Sheila handles the evening meal, usually by means of a take-out menu.

Cindy has just asked for a second helping of taco salad, which you'll gratefully accept as a "win," when you hear Sheila and Stephanie come in the front door. Stephanie walks right past you obliviously, her iPod so loud you can hear the buzzing of the music from the earphones. So you turn to face Sheila, hanging up her coat.

"So," you ask, "what did the doctor say?"

She doesn't even turn from the closet. "Oh, everything's fine."

"What exactly was this visit all about, anyway?"

"Nothing." She slams the closet door. "Girl problems. Don't bother yourself with it."

"If anything is wrong, I'd like to …"

"I took care of it, Wayne. End of story." She brushes past you, on her way to the kitchen. "What's for dinner?"

"Taco salad."

"Barf. Maybe I'll order some Pad Thai." She stops to bellow up the stairs. "STEPHANIE! YOU WANT SOME *PAD THAI?*"

You hear Stephanie's muffled "What?" from upstairs and Cindy's "I want *Pad Thai*!" from the kitchen, and you feel your newfound fame as Top Taco Salad Chef crumbling. There is an unreal moment when it seems that your entire household has become a map of secrets and lies … maybe you're even standing in the wrong house.

Reality is reinforced when you hear your phone chirping in its belt holster. Pulling it out, you see that it's your mother calling.

"Wayne," she says, "could you come over, please? It's important."

"Okay," you answer, glad for a chance to get out of the house before the Pad Thai arrives. "I'll be there shortly."

"Tonight, right?"

"Yes, Mama, tonight. Right now."

The radio is off as you drive to your mother's house. You want no distractions, but you're also not having much luck putting your thoughts in order. There's been distance between you and Sheila lately, sure, but it seems to be flaring up into outright hostility these days. You're angry with yourself for not insisting on being told why your oldest daughter was taken to the doctor. What if it really was just a female plumbing thing you couldn't relate to, anyway? Well, that just makes you angrier, and then you're mad at yourself over getting so angry. *Argh.*

You could use a beer at Fatso's right now. It's kinda on the way home … But no. If that little run-in with the law last week taught you anything, it's that Fatso's is not necessarily your friend. Besides, your mom needs you now. It didn't sound like it was for a leaky faucet or anything like that.

You pull up in the driveway of her modest bungalow, and rap at the front door before letting yourself in with your key. "Mom?"

"In the back, Wayne. Come on back."

You weave your way through a labyrinth of boxes and canvases. Mom got back into her painting in a big way the last few years. We moved most of the furniture into the front rooms to clear out a studio in back. She's sitting in her easy chair in that room now, looking at a half-finished portrait of an older woman, looking a bit forlorn, with a shock of gray hair. It's a self-portrait, and it's not bad.

"What's up?"

"I want you to look in my mouth," she says, rising.

"I ... beg your pardon?"

"Look in my mouth. There's a weird spot in there." She hands you one of those tiny LCD flashlights you buy in convenience stores for key chains. She opens her mouth and sticks her tongue out. "*Hee? Ah ya bakka yung?*"

You switch on the flashlight and look. "Yeah, I see something on the back of your tongue. The white spot, right?"

She closes her mouth. "Right."

"How long has it been there?"

"A few weeks. I thought it might go away when I first felt it, but it hasn't. If anything, it's gotten bigger."

"You think we should have it looked at?"

"I don't know. What do you think?"

You sigh. It doesn't look particularly threatening, but you haven't been good with decisions lately. Mom smoked heavily back in the old days, quitting only after your father died of cancer. She has a particular horror of the disease, and at the age of 80, you're not sure how it would affect her if she found out she had cancer.

On the side table next to her easy chair you see an open bottle of Canadian Mist, some amber residue still wetly gleaming on the bottom of a nearby glass. Yeah, she's worried already. For a moment, you feel like sharing a drink with her, then angrily shove the thought away. *What the hell.* You don't like these cravings for alcohol you've been having lately. Especially not now, when your mother needs you, sane and unimpaired.

You figure it's worth having Dr. Hernandez take a look at the spot. Then you visualize Sheila's face when you tell her you are taking your mother to the doctor for a tongue exam. *This will surely lead to an argument.* Sheila will probably call the spot, "A simple case of too many jalapenos." That's when the fight will start and it probably won't end well. *Is this spot worth another war at home?*

TURN TO PAGE 198 TO MAKE AN APPOINTMENT FOR THE SPOT.
TURN TO PAGE 199 TO IGNORE THE SPOT.

# Wayne: Make an Appointment

REALLY, YOU DON'T KNOW WHY YOU TOLD SHEILA. "A *spot* on her tongue? Is that all? Good Lord, Wayne, it's always something with your family. A Brewster can't stop thinking they're falling apart. I am sure you got that from your mother."

"I just think it should be looked at, is all."

"Do you know if her Medicare will even take care of it?"

You think maybe it will, but then again, maybe it won't. It doesn't matter. This is your mother you're talking about. You pull out your phone and find Dr. Hernandez' name on the contact list. You'll call first thing in the morning.

TURN TO PAGE 201 TO CONTINUE.

# Wayne: No Appointment

REALLY, YOU DON'T KNOW WHY YOU TOLD SHEILA. "A *spot* on her tongue? Is that all? Good Lord, Wayne, it's always something with your family. A Brewster can't stop thinking they're falling apart. I am sure you got that from your mother."

"So, you don't think it should at least be looked at by physician?"

Sheila's finger is wagging. "I think you come from a long line of hypochondriacs and she just bit her tongue or something like that. You want her to see a doctor over a tongue bite? Does Medicare cover self-inflicted wounds?" Sheila chuckles. "How much is the copayment for hypochondria?"

Well, that's that, you think, the almighty bottom line rationale delivered as an insult. You feel your blood pressure rise, then take a deep breath. *Hold it … hold it …* You exhale slowly. There, that's better. Maybe Sheila's right. Maybe the spot is nothing but a cold sore. Maybe you and your mother are getting yourselves worked up over nothing. You decide it's not worth arguing over.

"You're right, sweetie. It probably is nothing."

* * * *

Hours later, lying in bed, you're staring at the ceiling. If it was nothing, would it still be nagging at you like this? Would you be losing sleep over it? You glance over to Sheila, peacefully sleeping on her side of the bed, her back to you. No comfort there — she might as well be sleeping in another country, for all the closeness you feel with her tonight.

Finally, you decide, it doesn't matter. It doesn't matter if the spot is nothing. It doesn't matter if Sheila thinks it's hypochondria. It doesn't matter what Sheila thinks or says at all. This is your mother you're talking about. You decide that you are going to phone Dr. Hernandez' first thing in the morning and make an appointment, anyway.

Then sleep comes down like a warm bath.

TURN TO PAGE 201 TO CONTINUE.

# Wayne: Mom's Appointment

EXAM ROOM FOUR. Mom likes the woodpecker watercolor, too. To pass the time, you're about to talk to her about your possible return to watercolors when there's a rap at the door. Julie walks in.

There is the usual "Hello," the exchanging of pleasantries, the taking of vital signs. After Julie finishes scribbling in her folder, she looks up at your mom. "Well, Gloria, your vital signs are perfect, as usual. So what's going on with you today?"

There it is. The question. Waiting for this has had you increasingly nervous all morning, certain it is either going to provoke terrible news or Julie rolling her eyes at you. "She has a sore in her mouth!" you blurt out, almost immediately embarrassed.

Julie doesn't even look in your direction. "Can you show me where it is?"

"It's on her tongue," you say, and then decide to shut up as your mother dutifully sticks out her tongue.

"Hmm," Julie says, noncommittally. "How long has that been there?"

"A few weeks," your mother replies, before you can. "I kept hoping it would go away, you know? It hasn't."

"Has it changed at all in that time?"

"I think it's gotten bigger. I'm almost sure of it."

"Hmm." Julie writes something in the folder.

A bad feeling creeps over you.

"Is it bad?" your mother asks. "You know, I used to be quite a smoker in my younger days."

"You stopped, right? That's the best thing you could have done." Julie stands. "Dr. Hernandez needs to have a look at this, okay? I'll be right back with him."

Julie steps into the hall, and you follow her. She closes the exam room door, looking at you as if she knows what you're going to say. Heck, she probably does, she's a smart girl.

You can't keep it in. "It's cancer, isn't it?" Then the words start coming out in a flood, as if they were just waiting for you to say the C-word before they could come charging out. "Look, if it is, please don't tell her. My dad died of cancer back in '97 and now she's like, deathly afraid of it. She keeps talking about all her friends who are dying of it and keeps saying stuff like, 'Knock on wood, that won't be me — I couldn't handle it.' I don't think she could, either."

Julie waits, uncertain if you're finished. You don't feel that you are. "Look, I know I'm a worrier. I know I'm kind of a laughing stock around my house for that. For all I know, I am around here, too." She opens her mouth to disagree, but you press on. "I come by that honestly, my mom's the same way. I'm afraid if we tell her she has a serious disease, it will worry her to death. I mean *to death*."

Julie puts her hand on your arm, reassuringly. "Wayne, I can't tell anything just by looking at it. Neither can Dr. Hernandez, but he'll be able to say if we should go to the next step. Tell him your concerns, certainly, but let's not panic, okay?"

You sigh, and you do feel a little bit of the weight fall from your shoulders. "Okay." You take a deep breadth. "As I told you, I'm a worrier."

"Yes, you are." Julie smiles. "I'll go get the doctor now."

Dr. Hernandez comes into the exam room and listens to your mother. You feel a little talked out at the moment. She does the stick-out-your-tongue-and-say-ah bit. Finally he says, "I think you were right to come in guys. I'm concerned enough that I think we need a biopsy. If it's nothing serious, at least we'll know what we're not dealing with, right? Do you know an oral surgeon?"

Your mother shakes her head. "Not since I got the choppers back in the 80s. He's probably dead by now. Heck, he was ancient then."

"I'll have Lily call Dr. Nguyen and see who she recommends. It would be good if we can get you worked in soon. It might take a week before we know anything. Okay?"

Gloria is smiling. She's always liked Hernandez, and the doc seems to have a soft spot for her.

"Okay then," he rises. "You can probably go out in the waiting room while Lily calls her office. New magazines came yesterday."

You follow Hernandez into the hall, hoping to give him a pared-down version of the speech you gave Julie. Hopefully, one that doesn't make you sound like a complete babbling idiot. It's just that you're certain that bad news would absolutely kill your mother. You're certain of it.

TURN TO PAGE 204 TO CONTINUE.

# Wayne: The Affair

THERE IS AN OLD COPING MECHANISM your mother told you about once. She called it, "thinking that there are many people worse off than me," and then considering the plight of those poor folks and how lucky you are in comparison. In your experience, it's never worked.

Last night, things finally came to a head with Sheila. You knew they eventually would. The long days away from home on business — sure, that was bothersome. It's the change in the way she treats you that's the worst. Most of the time she seems to be dismissing you like an irritating servant. When she's not, in the increasingly rare times she's intimate with you, there seems to be an underlying tension, a feeling of a duty best over with quickly. Maybe it's … a feeling of guilt?

So last night you were tired and stressed out enough to be blunt. You asked her if there was another man, even though you felt like you were in some brightly lit daytime soap opera when you did it. You're still mulling over her reaction in your head. *Was it too quick, like she'd been expecting the question? Was she too slow, as if surprised?* You're not sure, because what sticks with you is that she laughed. Then she said something like "You wish," or maybe it was "I wish." But, she laughed. She laughed at the idea. She laughed at you.

Then, this morning, your Jeep suddenly stops working on the freeway on your way into work. You manage to get it to the shoulder, spend some time looking under the hood, like you knew what you're looking at, then call AAA and wait for the wrecker. They're surprisingly sympathetic when you call into work. That's a pleasant surprise. What's more surprising, or maybe worrisome, is that they also give you the day off.

You ride in the wrecker to Freddy's Auto Clinic, where you're becoming a regular. Freddy says he'll give you a call when he figures it out. Do you need a ride? "No," you reply, "the doctor says I need the exercise and it's less than a mile to my house. I'll walk."

You feel a little better as you turn down your block. Your mind isn't any clearer, but the time alone has let you sort some things out, get a better grip on what's important. Prioritizing, they called it in those meetings at work.

Sheila's car is in the driveway. Not too uncommon. *Well, you think, we may be getting one of those priorities out of the way right now. Sheila needs to know what I'm feeling.* You two need to talk about your marriage, and without the kids around, right now is the time to do it.

You let yourself in the front door. Sheila is in back, in the kitchen, banging around, slamming cabinet doors. *Great, she's already in a mood.* You're about to call out to her when your hear her phone's ring tone, that old opening stanza from "Lola" — and to think, you used to like The Kinks — and you stop. Something, some lingering suspicion, tells you to stop and listen.

The ring tone ends, and there is the briefest of pauses before Sheila snarls into the phone, "Don't you say, 'Hi' to me, Julie! *You have some nerve!*"

*Julie? The nurse at Hernandez' office? Why is she —*

"You *don't know* what's upsetting me, Julie? *Ha!* You should have heard the conversation I had with my husband last night! He knows I've been sleeping with Armand, Julie, and you know what? It had to be *you*, Julie! Only you and Dr. Hernandez knew, and *he* would *never* betray my confidence. It. Had. To. Be. You!"

*Wait, Armand? Her boss?* Sheila's reaction to your question last night now makes a lot more sense now. *Many* things now make a lot more sense.

Sheila is fuming. "I'm calling my good friend Carla right now, Julie. You know she's an attorney, don't you? Not to mention your boss' wife? I suggest you start looking for a new job, Julie. And an attorney. *Because, you're going to need one!*" Sheila probably wishes she wasn't on her iPhone so she could hang up by slamming it down on the cradle for emphasis.

What were you thinking a few minutes ago? *We may be getting one of those priorities out of the way right now. Sheila needs to know how you're feeling? You two need to talk about your marriage?* You step into the kitchen. The look on Sheila's face is priceless.

"So," you say. "Am I going to need an attorney, too?"

Turn to page 251 to read Clinical Ethics.

# Julie: My Phone

YOU PULL YOUR PHONE FROM YOUR BACK POCKET and see the familiar NUMBER BLOCKED indicator. Another telemarketer. You click the OFF button and tend to Walter's vital signs.

"How ya doing?" you ask Walter.

"I'm okay."

His vitals say he's okay, but you know something is up with that boy. You have been treating his family ever since you started working for Dr. Hernandez, more than 3 years. You know his mother, Sheila; his two sisters, Cindy and Stephanie; his grandmothers, Gloria and Barbara Ann; and of course his dad, Wayne.

"Are you sure you're okay? You know your dad is on the phone with your mom. He can't hear anything you say. I sure as heck won't tell him anything," you probe.

"Thanks, but I'm fine ... really ... I just didn't pay attention to how much I was drinking. It's an honest mistake. No big deal." Walter's voice is remarkably calm.

You decide to leave it at that but take note to keep an eye on him when you see him around campus. You give Walter a consent form to sign, after which Dr. Hernandez examines his wounds and asks a few questions to check for a concussion. Following the examination you help clean and dress Walter's wounds.

Wayne, who has been on the phone the entire time, finally hangs up. "Everyone sends their love. I told them you were banged up but otherwise well," he tells Walter.

Dr. Hernandez asks you to prepare a tetanus shot and to get a couple of vials for some blood work. In a soft voice, he confides, "He'll be fine, Wayne. He's probably going to be sore as hell, but that's to be expected. Just keep an eye on him. If he starts getting dizzy or feels like vomiting give me a call."

You bring the hypodermic and blood vials to Hernandez. He pokes the needle into Walter's arm and injects the serum. "Your arm will probably be stiff, but it shouldn't last long. While I'm here, I'd like to do a blood workup to make sure your white count is okay and there is nothing going on internally."

"I'll pass on the blood work," Walter says.

"If you're worried about you blood alcohol level, you know HIPAA prevents me from giving out patient information to the police or anyone."

"I'm good. I just want to go home."

Hernandez doesn't push Walter any further. You wonder what's up with Walter. You think, maybe he has something to hide? Again you feel that something isn't quite right with the guy.

"Well, then, you are free to go. I can't find any evidence of a concussion, but it's not unheard of for symptoms to show up two or three days after a trauma. We're going to have to keep a close eye on you, son. You were very, very lucky tonight, Walter — but you're not going to feel very lucky for the next few days. How are you set for Ibuprofen?"

"I'm good. Got plenty of it."

"Then take as directed. And, either of you let me know if there's any change. Okay then. He's your problem now, Wayne."

You and Wayne help Walter down from the examination table and move him slowly toward the back door. You hear Hernandez' voice behind you, "Hopefully I won't see either of you for a while."

Wayne turns with a half smile on his face, "Sorry, Doc, I'll see you later this week. I have an appointment on Tuesday." Hernandez seems almost amused, but not at all surprised.

You watch Wayne and Walter hobble into the night air. You wonder what the ride home will be like. His father didn't seem to be too upset, and that bothers you a little. You chuckle as you think about the Brewsters. They're a sweet little family — each one unique and a bit quirky in their own special way.

Before you leave you confide in Dr. Hernandez. "I am beginning to think that the Brewsters may be frequent flyers. Mr. Brewster, in particular, seems to come in with no good reason."

Hernandez nods in agreement. "Yes, but at the same time, there may be something going on that we are not aware of just yet. In my experience, underlying causes reveal themselves, eventually. We want to do everything medically necessary for the family. That sometimes includes helping them feel comfortable about their health. Sometimes that means indulging them, even when we suspect a bit of hypochondria."

You feel good about Dr. Hernandez' attitude. Treating the whole patient has always been at the center of his practice. You remind yourself that part of being a successful clinician is dealing with families exactly like the Brewsters.

You and Hernandez close up shop. He tells you how much he appreciates your help and bids you a good night. You get in your car and enjoy a slow ride home. It's one of the benefits of living around Barton Creek. The night skies are spectacular. Every night Mother Nature takes her brush and composes a new work of art for all to see. Tonight is no different.

TURN TO PAGE 211 TO CONTINUE.

# Julie: DEXA Scan

TUESDAY ROLLS AROUND. It's another busy day in the clinic. Wayne Brewster arrives, early as usual. You escort him to the examination room.

"How's my blood pressure? I don't have a temperature do, I?"

Remembering your conversation with Dr. Hernandez, you try to comfort Mr. Brewster. "Your vital signs are perfect. Your blood pressure is like a teenager's."

Wayne smiles at your comment.

"How is your son doing?"

"Walter? Doing well, thank you for asking." Walter sighs in relief. "He's moving around more, and going back to classes. Though I guess that's more due to his mother driving him nuts than anything."

"Good. I am glad to hear that. Now, what brings you in to us today?"

You listen attentively as Wayne explains the reason for his visit. He starts by reminding you that he has just turned 50 and has begun to have aches and pains. He keeps thinking about his 74 year-old mother-in-law's recent diagnosis of osteoporosis that was confirmed by a DEXA scan. He wonders if he might have osteoporosis too. He wants a DEXA scan to see if he has osteoporosis. If he does, he wants to start taking medication right away. Even if he doesn't, he wants to know if taking medication now might prevent osteoporosis in the future.

You know there is no way Wayne has osteoporosis. But, Hernandez said, you sometimes have to indulge a patient's concerns. You excuse yourself and tell Mr. Brewster that the doctor will be right in.

Outside the exam room you explain the situation to Dr. Hernandez. He agrees that Mr. Brewster does not have osteoporosis. However, his experience with Wayne Brewster is that he will expect a test. At times it's okay to order a few labs to put a patient at ease, but ordering an expensive scan without reasonable cause is "problematic." You're sure this isn't the first time, nor will it be the last, that this issue comes up. In fact, Dr. Hernandez had the following excerpt from a modern version of the Hippocratic oath framed on the wall in the hallway for this very reason. It reads,

> I will remember that I do not treat a fever
> chart, a cancerous growth, but a sick
> human being, whose illness may affect the
> person's family and economic stability. My
> responsibility includes these related
> problems, if I am to care adequately for
> the sick.

Lily, Hernandez' office manager, passes by and overhears your conversation. She reminds you and Dr. Hernandez that the clinic needs to bill six scans per month to pay for the new machine. She also notes that many insurance companies will pay for the test, especially if there are risk factors.

"Is the scanner occupied?" Hernandez asks.

You check with the technician who tells you the scanner is available.

Hernandez has a frown on his face.

Wayne Brewster opens the door and pokes his head out through the opening. He has a worried look on his face.

"Hello, Wayne." Hernandez reaches over to shake his hand. "Hang on a second. I'll be right in."

Hernandez turns to you. "I really don't know about this."

TURN TO PAGE 214 TO SUGGEST A DEXA SCAN.

TURN TO PAGE 217 TO SUGGEST A DEXA SCAN IS NOT WARRANTED.

# Julie: Order Scan

"WE TREAT THE PATIENT'S CONCERNS as well as the symptoms, right?" you say.

"Okay, Julie. Wait here." Dr. Hernandez walks into the exam room. He returns ten minutes later and hands you an order for a scan. "I told him I didn't think he had osteoporosis and that there was no medical justification for ordering the test. He insisted, as we expected. So, here is the order."

A few hours later, you escort Wayne Brewster to the scanner room.

"It doesn't hurt or anything, does it?" he asks. "Say, do you know how much the test is if the insurance doesn't cover it? I mean, you know, with Walter in medical school, and Stephanie rapidly approaching the age where I have to think about college for her, too."

"Kids are expensive, so they tell me." You offer to have Lily go over the financial details with him before the scan. He tells you not to bother because his peace of mind is more important to him than the money. You introduce him to the technician. Before you leave you tell Wayne, "It was nice seeing you again. Please tell Walter 'Hello' for me and give my regards to the rest of your family." You shut the door and head for the stack of patient charts that are piling up. It's time to leave the Brewsters behind and move on to the next patient.

TURN TO PAGE 215 TO CONTINUE.

# Julie: Insurance Won't Pay

THREE WEEKS LATER, you overhear a phone conversation between Lily, the clinic manager, and Wayne's insurance carrier. The company does not feel the DEXA scan was indicated and is refusing to pay for the test.

Lily is a bit irritated. She looks at you. "I told you it was pointless to run the Brewster scan through his insurance."

"That's not the way I remember it," you remind her. Seems like this is why you hate the business side of this business. Your concern has always been patient care, as you feel it should be. You are fairly oblivious to the financial side of things and are comfortable with that. As a soon-to-be nurse practitioner you made a commitment to treat all patients regardless of their ability to pay. You can't figure out people like Lily and those who run insurance companies.

A few minutes later your cell phone rings. It's Wayne Brewster. "Julie, I hate to be a bother, but I think my blood pressure is out of control. Lily just told me that my insurance will not pay for the scan and I have to fork over $350. Sheila is going to hit the roof. I feel awful. I use to be able to handle things like this, but lately I just feel so out of control. I know my reaction doesn't make any sense but maybe Dr. Hernandez can call in some anti anxiety something-or-other for me to help me through this."

You pause for a moment and try to calm him down. "Mr. Brewster, I completely understand your concern. Have you taken your blood pressure? ... You have? ... 120 over 75? Why, that's perfect. I know you are upset over the bill. That's understandable. I would be too. No, I don't have anything to do with the billing, but I bet Lily will be happy to work with you. Why don't you speak with her again and see what she can do?"

Your comments seem to calm him down. He thanks you for your help. He hangs up. Five minutes later your phone rings again. It's Wayne Brewster. It looks like it's going to be one of those days. "Hello Mr. Brewster. How may I help you? Yes … oh … uh huh … really?"

TURN TO PAGE 211 TO TRY AGAIN.

# Julie: No DEXA Scan

"THE TEST DOESN'T SEEM INDICATED TO ME," you say.

"Yeah, I agree. Wait here." Dr. Hernandez walks into the exam room. He returns ten minutes later. "I told him we didn't think he had osteoporosis and that there was no medical justification for ordering the test. He insisted, as we expected. I suggested that he take calcium and promised him that I would monitor him closely for signs of osteoporosis. I think that made him feel better."

You are pleased with the outcome. You return to the exam room with some instructions for Wayne, then escort him to Lily for checkout. Before you leave, you tell him, "It was nice seeing you again. Please tell Water 'Hello' for me and give my regards to the rest of your family."

You head for the stack of patient charts that are piling up. It's time to leave the Brewsters behind and move on to the next patient.

TURN TO PAGE 218 TO CONTINUE.

# Julie: Wayne's Letter

IT HAS BEEN ONE MONTH since the DEXA scan episode. You are beginning to view the life of a Brewster as a never ending saga in the Book of Family Medicine. Today, Mr. Brewster is in the clinic again. He called earlier in the morning saying that he had to be seen "right away." You call him in from the waiting room and escort him to the exam room. As you take his vitals, you notice that he has dropped a little weight. You make a note of it in his chart.

"So, what can we do for you today, Mr. Brewster?"

"Thank you for always listening to me, Julie, I appreciate it." You notice Wayne seems to be avoiding eye contact.

"Well," he continues, "I need a written excuse for my employer — just a simple letter stating, 'Wayne had a bad and potentially contagious GI problem.'"

You flip through his chart. "Have we seen you for a GI problem recently?"

"Well, no, not exactly," he says, "My daughter, Cindy, is the one who was sick, but I had to stay home with her and post office policies only excuse absences due to employee illness. They don't give me a break if I have to stay home and care for a family member. I can't afford anymore absences."

Wayne seems desperate. It seems as if his anxiety over dealing with the normal stresses of life could be managed somehow, but you are not sure what to do about it. You feel you can't just flat out lie for the guy. You are grateful you are not the one who has to decide how to handle this. You tell him you are going to talk to Dr. Hernandez and that you will be right back.

You exit the exam room in search of Hernandez.

TURN TO PAGE 222 TO SUGGEST HERNANDEZ WRITE THE LETTER.

TURN TO PAGE 220 TO SUGGEST HERNANDEZ NOT WRITE THE LETTER.

# Julie: No Letter

YOU WAIT A FEW MOMENTS for Dr. Hernandez in the hallway. When he exits an exam room you confront him with the dilemma. You explain the situation about Wayne and the letter. You say that you're not comfortable with the situation.

Hernandez looks troubled. He outlines the ethical considerations to you, but agrees that Wayne seems to be in a stressful situation. On the other hand, he also says writing the letter may do more harm than good, as it could be encouraging irresponsible behavior.

You and Hernandez enter the exam room. Hernandez tells Wayne Brewster that he cannot, in good conscience, write the excuse letter. Instead, Hernandez suggests Walter be honest with his boss, who might have kids himself and understand the situation. Wayne appears deeply troubled. Dr. Hernandez pats him on the back and leaves the room.

"I'm sorry, Mr. Brewster. I'm sure your boss will be understanding."

"You don't know my boss." Wayne stands up. You open the door and watch Wayne Brewster drag himself through the lobby to the parking lot.

TURN TO PAGE 221 TO CONTINUE.

# Julie: Wayne's Reprimand

WAYNE BREWSTER SITS IN HIS JEEP. He's sat there for 15 minutes doing nothing but thinking — thinking that he had to do it. He had to go in and tell his supervisor, Joe, that he wasn't sick; that he was staying home with a sick daughter.

It didn't go over well; even worse than Wayne had expected. He wasn't expecting an understanding smile and a, "There, there, that's okay!" But, *good grief.* It started with the docking of a week's pay. That was sort of expected. Then there was the going on and on, the doubting of Wayne's commitment to his job, the level of trust that could no longer be given him, that this was going to go on his record, and that he could forget about "that raise," let alone a promotion …

*Hell,* Wayne thinks. *I have worked at the Post Office for 25 years, and I am never going to be anything more than a mail carrier.* He sighs heavily and starts the Jeep's engine. It's time for a pit stop at Fatso's Bar, a ritual he didn't indulge while Cindy was sick — he'd been too busy trying to play catch-up and come back to work and make a good impression. *All for nothing.*

The beer is cold but the camaraderie warm. He talks with the other regulars. Everybody's got complaints about their jobs, and they try to outdo each other with their Boss From Hell stories. Wayne feels better. A little less alone. A little less hurt. He leaves to go home around 1:30 AM.

TURN TO PAGE 223 TO CONTINUE.

# Julie: Write Letter

YOU WAIT A FEW MOMENTS for Dr. Hernandez in the hallway. When he exits an exam room you confront him with the dilemma. You explain the situation about Wayne and the letter. You tell Hernandez that you are sympathetic with his problem.

Hernandez looks troubled. He outlines the ethical considerations to you, but agrees that Wayne seems to be in a stressful situation. He agrees to write the letter. "Well, I guess it's not hurting anyone. Could you draft something for me to sign?"

You do as Dr. Hernandez says. Then you start having doubts if it is the right thing to do. You feel that by writing the letter, Dr. Hernandez may be encouraging irresponsible behavior. *Maybe Dr. Hernandez should suggest Mr. Brewster seek counseling instead?*

What's done is done. You return to the exam room and hand Mr. Brewster the letter. He seems relieved. On the other hand, you are sure it is only a matter of time before he is back with another problem.

TURN TO PAGE 223 TO CONTINUE.

# Julie: Stephanie's Appointment

IT HAS BEEN A COUPLE OF WEEKS since you last saw a member of the Brewster family. Today, you will see Stephanie Brewster, Wayne's oldest daughter. She's here with her mother, Sheila Brewster. Apparently Stephanie has come in to discuss her concerns about her acne. You enter the room and welcome both mother and daughter.

"Well, hello there. My goodness, Stephanie, I can't believe how much you have grown," you say with genuine enthusiasm. You have been her nurse for several years now. You have watched her grow from a kid into a beautiful teenager, and like most her age, she regularly wears clothing that reveals her navel and a significant amount of cleavage. In your opinion, it isn't acne she should be worrying about.

You take her vital signs and notice that she has grown two inches and completely lost her baby fat. Amazing, you think. There's nothing quite like watching a child grow. It's truly a miracle.

You begin with your standard opening line. "So, what brings you in today?"

Before Stephanie can speak, Sheila jumps in. "It's her acne. We just can't seem to get it under control."

You look closely at Stephanie face, but don't think her acne is bad at all.

"Hmm. Well, Dr. Hernandez will be right in. It's good to see you both." You leave the room and wait for Dr. Hernandez in the corridor. A minute later, you follow Dr. Hernandez back into the room. He delivered Stephanie and has been her family physician ever since.

"Stephanie. Sheila. Great to see you both." Hernandez shakes Stephanie's hand. "Man, it's hard to believe that you are almost sixteen years old now. How old does that make me? Don't answer. I already know: Ancient." Hernandez pronounces Stephanie's acne well under control. In cases like this, he would recommend using Cetaphil as a basic cleanser and benzoyl peroxide every other day. He also realizes that, now that Stephanie is turning sixteen, she might want to talk about boys. He feels he has a good relationship with her and that she trusts him.

Hernandez gives you a knowing smile. You pick up the hint and suggest making Sheila some tea. You lead her out of the exam room toward the coffee bar.

Back in the exam room Hernandez asks Stephanie about sexual activity. Stephanie says, "My boyfriend and I have kissed, but I haven't done anything more than that. Both of us want to wait to have sex until we're married. I know my mother thinks that me and my friends are all 'doing it,' but we're not."

Hernandez is glad to hear that. He tells her she is always free to call and speak with him any time she likes.

You reenter the room. Hernandez hands you the chart and tells Stephanie and her mother goodbye. As he leaves, Sheila walks back into the hall, grabs Hernandez' sleeve and closes the exam room door behind her. You are standing close enough on the other side of the door to overhear their conversation.

Sheila tells Hernandez that she read in Cosmopolitan there are now birth control pills that are also "good for acne." Hernandez says nothing. "My husband Wayne is such a prude. I don't think he could ever accept that our daughter would have sex before getting married. I've been a young woman before and I understand the pressures on Stephanie. I know the most important thing is to prevent her from getting pregnant and ruining her life. I went to see our old friend, Dr. Small about this, but he refused to prescribe The Pill because he didn't believe in birth control. I think he was Catholic. So, please Dr. Hernandez, just write the prescription for the birth control pills that help with acne, and don't tell Stephanie what they are. Above all, don't tell Wayne. He's going downhill fast as it is — did you know he almost lost his driver's license after a DWI last month? His boss wrote him up and docked his pay for taking off when Cindy was sick. He feels he has ruined his chance for a promotion. He was so upset he went out and got drunk."

You open the door as Stephanie enters the hall. Dr. Hernandez instructs mother and daughter to wait in the lobby. He turns to you and starts to talk, but you say, "I heard."

Hernandez looks confused. "You're a bit closer to this demographic than I am, Julie. The daughter says she is not sexually active. I believe her. Her mother wants her to take birth control for her protection. I can understand that. But, Sheila wants me to prescribe it under the guise that it's for her daughter's acne. On that, I'm not so sure. What's your take?"

TURN TO PAGE 226 TO SUGGEST PRESCRIBING BIRTH CONTROL PILLS.
TURN TO PAGE 228 TO SUGGEST NOT PRESCRIBING BIRTH CONTROL PILLS.

# Julie: Birth Control

YOU ARE PLEASED that Dr. Hernandez is confiding in you. Having worked at a children's hospital you know the subject of teen pregnancy all too well. You believe it's better to be safe than sorry. You know how difficult it is for young women to deal with unwanted pregnancies. You are well aware that teen pregnancy has become a serious public health issue.

You voice your opinion. "If Stephanie asked you to prescribe The Pill for her without her mother knowing, then it would be easy to say, 'yes,' because as I understand it, the law does not require the parent of someone Stephanie's age be notified."

Hernandez nods.

"I also believe that while Stephanie says that she is not sexually active, it could change at any time, and if it does, it would be good for her to be protected. However, I believe it would be unethical to prescribe a medication for anyone without letting them know what they were taking. Don't we need Stephanie's informed consent?

Hernandez nods.

"So, I would tell Mrs. Brewster that you will prescribe The Pill, but not secretly. You will let Stephanie make the final decision. You will let her know that you will be prescribing a birth control pill that has residual benefits for the control of acne, but also possible side effects. In so doing, you maintain your integrity, Mrs. Brewster is pacified, and Stephanie can continue to practice abstinence with a clear complexion.

Hernandez smiles.

"I also believe that it would be good to encourage Mrs. Brewster to tell her husband about the pills."

"Ms. Spade, I applaud your reasoning. Bring the Brewsters to my office."

You escort mother and daughter to Hernandez' office and close the door. You pause for a moment before greeting the next patient. It looks like a win-win situation. You feel good about your assessment and hope it works out for everyone.

TURN TO PAGE 230 TO CONTINUE.

# Julie: No Birth Control

YOU ARE PLEASED that Dr. Hernandez is confiding in you. Having worked at a children's hospital you know the subject of teen pregnancy all too well. You know how difficult it is for young women to deal with unwanted pregnancies. You are well aware that teen pregnancy has become a serious public health issue.

You voice your opinion. "I believe we have to treat the patient, Dr. Hernandez. Regardless of what Mrs. Brewster wants for her daughter, Mom is not the patient. There is no way that we could ever justify lying to Stephanie, or any patient for that matter."

Hernandez looks at you quizzically.

"I mean, there may be a special circumstance where one might consider lying, perhaps when a patient's cognitive abilities are impaired to the point they can't make decisions for themselves. However, that isn't the case with Stephanie. She didn't request The Pill, so she should not be taking it. Anyway, we would need Stephanie's informed consent, wouldn't we?

Hernandez nods.

"You've started talking about these issues with Stephanie. You could begin a conversation with her mom and dad. Help them to understand and trust their daughter. Encourage them to engage her in conversations about the subject so that, when she becomes sexually active, she will be able to talk about her concerns and take appropriate precautions. I'd say the worse thing for a young person during this time in his or her life is to feel isolated."

Dr. Hernandez closes his eyes, then shakes his head. He gives you a look you have never seen before. "Ms. Spade, I applaud your reasoning and am impressed with your logic. You are going to be a wonderful practitioner. Please, bring the Brewsters to my office. I won't be writing any prescriptions for Stephanie today."

TURN TO PAGE 230 TO CONTINUE.

# Julie: Gloria's Appointment

IT'S SEVEN O'CLOCK IN THE MORNING. The clinic is quiet. You are a little tired, but that is to be expected. Working and going to school is stressful, but certainly worth it. You have loved being a nurse for all the right reasons. It has been the perfect blend of caring and science. That feels good. You are looking forward to the expanded horizons your advanced degree will bring.

Your first patient for the day is Gloria Brewster, Wayne's mother. Gloria is a sweet, unassuming octogenarian who is cognitively fit and in good health. Today, her sometimes histrionic son (as you think of him) has brought her in. You enter the examination room and welcome the two of them. After a bit of polite conversation and the taking of vital signs you get down to business. "You appear to be in good shape, Mrs. Brewster. How can we help you?"

"Call me Gloria, please. I'm not formal."

She tries to continue, but Wayne cuts her off. "She has a sore in her mouth."

"Can you show me where it is, Gloria?"

Wayne interrupts again. "I'll show you. Mom, open your mouth."

Wayne is clearly in the way. You know it is common for family members to speak for one another. Parents do it for children and when their children become the caregiver, roles reverse. However, you believe it is important for patients to speak for themselves whenever possible. You listen to Mr. Brewster, but continue to direct your inquiry to his mother.

She opens her mouth. You see a small white spot on the back of her tongue. Gloria says it's been there a few weeks and assumed it would go away. However, it has not gone away. It seems to be getting bigger. Upon examination you determine the lesion is suspicious. You are fairly certain that Dr. Hernandez will want to refer her to a specialist.

"Is it bad?" Gloria's voice quivers with sweetness and worry. "You know, I use to be a heavy smoker in my younger days."

You sense her concern. "You stopped, right? That's the best thing you could have done." You are puzzled, and it probably shows. "Hmm … It may be nothing. I am going to go get Dr. Hernandez so we can take care of this right away. I'll be right back."

Wayne follows you into the hall. "It's cancer, isn't it. Look, if it is, please don't tell her. My dad died of cancer back in '97 and now Mom's like, deathly afraid of it. She keeps talking about all her friends who are dying of it and keeps saying stuff like, 'Knock on wood, that won't be me — I couldn't handle it.' I don't think she could, either."

You place a reassuring hand on Wayne's shoulder and tell him he should speak with Dr. Hernandez about his concerns. That seems to calm him down and he returns to the room.

In the hall, you discuss the situation with Dr. Hernandez. He decides, based on your assessment, that a biopsy is warranted. He says he will call Dr. Nguyen from the dental school for a consult. Wayne must have heard Hernandez' voice in the hall. He comes out of the room and closes the door behind him. As you make your way down the corridor to the next exam room, you see Wayne has cornered Dr. Hernandez in the hall. Poor Wayne, you think. *He's such a worrywart. Still, he has a lot on his plate — an aging mother, a teenage daughter, an absent spouse, and a ten-year-old. Yep, Wayne Brewster is a bonafide, card-carrying member of the Sandwich Generation, feeling the squeeze.*

TURN TO PAGE 233 TO CONTINUE.

# Julie: Mrs. Jones' Pills

IT'S BEEN A LONG DAY AT THE CLINIC. Flu season is starting early this year. You had to work in more patients than usual. Both you and Dr. Hernandez are worn out. It's almost 5:30 PM when you and Hernandez debrief to discuss the patients you have seen in clinic today, including Gloria Brewster. After the debrief, Hernandez heads home. You switch the phone to the answering service, return a few calls, finish your progress notes and begin entering patients into the office EHR system.

In the front office, the receptionist is totaling up the afternoon's charges and preparing the bank deposit. Throughout your time in the office, you have noticed how competent and efficient the staff is. You are particularly impressed with Mrs. Jones. She gets along so well with the patients. You believe that Dr. Hernandez is fortunate to have a receptionist like her.

You hear her talking to a patient. "Yes, I know you need to see a doctor, but have you tried the county clinic. I know they take Medicaid … Yes, we use to take Medicaid, but we don't anymore. I *am* sorry. I wish I could help you … You're welcome … yes … yes … goodbye."

You are surprised to learn that the clinic doesn't take Medicaid anymore. You think about asking Hernandez about it tomorrow, but then remember that it is not your practice and, therefore, really not your business.

You let the EHR screen saver turn off the monitor. You close your eyes for a moment. You are so tired that, if you had a change of clothes, you just might consider sleeping on the couch in Hernandez' office. The faces of patients you have seen today float in front of you. You think about Mrs. Brewster. You hope she doesn't have cancer.

Just as your head becomes heavy and falls to your chest, Mrs. Jones peeks into the work room. "I'm sorry to interrupt you, Julie."

You slowly open your eyes. "Yes?"

"Well, I just got a call from home. The sitter says my son isn't feeling well and is complaining of a sore throat again. Sounds just like what he had before. Last time, Dr. Hernandez gave him one of those *Z-Paks* and his throat got better really quick. The drug rep was here today and left a few *Z-Pak* samples and … well … I was hoping, although I hate to ask …. But I don't want him to get sicker, and I know that your class schedule next week will keep you out of the clinic a lot, and I don't want to have to stay home when Dr. Hernandez' schedule is so packed. I mean, I need to be here in the office, not staying home with a sick child. So, I was hoping that maybe … I mean, I would very much appreciate it if you could give me one of the *Z-Paks*."

TURN TO PAGE 235 TO OFFER A Z-PAK TO MRS. JONES.
TURN TO PAGE 239 TO DENY A Z-PAK TO MRS. JONES.

# Julie: Z-Paks to Go

YOU THINK ABOUT MRS. JONE'S REQUEST. Under normal circumstances, antibiotics should not be prescribed without seeing a patient first. Of course, the truth is, you can't prescribe medication at all — period. It would be illegal. On the other hand, this is an employee who is just taking home a sample for a recurrent condition. Since the boy has taken it before, it seems reasonable to give it to him again. You don't see the harm in letting her take a sample. You're a nurse after all, and you know something about prescription medicine. "Sure, Mrs. Jones. Go ahead and take one."

"Thank you. I really appreciate it."

"I hope your son is feeling better real soon." You go back to the last few entries for the EHR.

TURN TO PAGE 236 TO CONTINUE.

# Julie: Dr. Hernandez Finds Out

THE NEXT DAY, Dr. Hernandez overhears Mrs. Jones thank you for the *Z-Pak*. She says, it was "just what the doctor ordered." Of course, it was not what the doctor ordered. Hernandez calls you into his office.

"Ms. Spates," he begins. The way he says your last name is a little disconcerting. You can feel your amygdala pounding in your temporal lobe. Your *fight-or-flight* response goes into overtime. You get the sinking feeling that this is not going to be fun. You take a seat and wait for … whatever.

"I'll get right to the point. You don't have a license to prescribe medication. I am the only one in this office who is qualified to do that. I don't care what you were thinking. I don't care what Mrs. Jones told you. You don't know anything about the patient. You know as well as I do that you shouldn't offer so much as an aspirin without reviewing a chart and assessing the patient, in person. Do I make myself clear?"

In all your years in nursing you have never had your face ripped off — till now. It feels horrible. Of course, you know he's right. You leave his office, tail between legs and try to move on with the rest of your day.

Later in the afternoon, another disturbing issue surfaces. This time you are more mad than upset. You overhear Lily mention sending "The Letter" to Mrs. Johnson, a sweet sixty-five year-old patient you had seen when she came in for an annual mammogram and Pap smear. You felt sorry for her because she had just lost her job and was feeling depressed.

When you ask Lily what "The Letter" is, she tells you it is the one the clinic sends dismissing patients for failure to pay their bill.

*That's horrible,* you think. You can't believe that they are dismissing poor Mrs. Johnson because she can't pay her bill. You go on with your day, but are bothered by this for the remainder of the morning.

During lunch you tentatively approach Hernandez sitting in his office. "Excuse me, Dr. Hernandez … but I'm a little concerned about something."

"What is it?" Hernandez speaks without looking up from the chart he is reviewing on his desk.

"I didn't realize that we did not take Medicaid."

"We had to stop." Hearing the disappointment in your voice, he puts down the chart. "I know it may seem harsh, but this *is* a business. I can't afford to pay for this practice without billing for services. I want to help as many people as I can, but I can't afford to participate in Medicaid anymore. The paperwork is horribly time consuming. The reimbursement rates have gotten so low that I am losing money on Medicaid patients. Lots of money. I have no choice."

You see Lily standing in the doorway, obviously overhearing much of the conversation from the hall. She adds, "And besides that, and I hate to say this, but that kind of patient just scares away all the paying customers."

You feel your blood pressure rise. "What about Mrs. Johnson? How can we drop her when the poor woman just lost her job, and along with that, her insurance?"

Lily continues, "We offered her a payment plan which she has yet to accept. Does a mechanic fix your car if you don't pay? Of course not. This is the same thing."

You bite your lip and nod your head, realizing this may not be the best time and place for your inner socialist to speak up.

TURN TO PAGE 233 TO START AGAIN.

# Julie: No Z-Pak

YOU THINK ABOUT MRS. JONES' REQUEST. Under normal circumstances one should not give antibiotics without seeing a patient first. Truth is, however, you can't prescribe anyway. You think, *this is an employee who is just taking home a sample for a recurrent condition. Since the boy has taken it before it seems perfectly safe to give it to him again.* But, you realize the law is the law. You politely apologize, "Mrs. Jones, I know it seems like a simple thing. But I am not licensed to prescribe medication. I hope you understand. If you want, I can call Dr. Hernandez on his mobile phone to see if it's okay."

Mrs. Jones appreciates your offer. "That would be great. Thanks, Julie."

TURN TO PAGE 240 TO CONTINUE.

# Julie: Sheila's Appointment

AGAIN, ONE OF THE BREWSTERS IS IN THE OFFICE. This week, it's Wayne's wife, Sheila, with an appointment for allergies.

"I just can't get any relief. I have tried nose sprays and over-the-counter allergy pills. Nothing works. I tried salt water in a neti pot — no relief. I tried sleeping with a humidifier and hovering over steaming pots of water with a towel over my head. Nothing is working. I am miserable."

You have never seen Sheila Brewster so stressed. Her level of anxiety does not seem appropriate for a nasal drip. You try to empathize with her. "I know how you feel. I'm allergic to pollen. It's horrible. Let me go get Dr. Hernandez. I am sure he will be able to help you."

You wait for Dr. Hernandez in the hall. As you and he walk down the corridor, you explain Mrs. Brewster's concerns. Back in the exam room, Sheila thanks you both for taking care of Walter after his "little fender bender" a few months ago. It is obvious that Mrs. Brewster is down playing the seriousness of her son's accident. *Why would she do that?* Again, you feel something isn't quite right about this clan.

Hernandez completes his examination and begins to present Sheila with a treatment plan. She interrupts him mid sentence and delivers an unexpected diatribe in one breath.

"You know, my job has been keeping me extremely busy lately. I'm not complaining, mind you. I'm very successful. I don't know if I told you that I have been promoted to a high level management position. I am not happy about the additional hours, but it goes with the territory. Wayne doesn't like it one bit. He is always complaining that I don't have enough time for the kids, but he could help them with their homework just as well as I can. Maybe he should become a full-time house spouse? All he does is complain that I don't give him enough attention. I think he's upset that I make more money than he does. Why can't I be nicer to his mother? I'm just sick of all his whining."

As she continues, the real reason for Sheila's visit becomes clear. Just like bringing in her daughter, Stephanie, ostensibly for acne but really for birth control, Sheila has an ulterior motive. She explains that on her last business trip, she enjoyed staying out at night and "talking" with Armand, her boss. Now she's worried that she might have some kind of infection, and she could give it to her husband. She's tried to avoid any situations that could lead to having sex with her husband, which isn't all that difficult, given the state of their marriage. If she continues making excuses, Wayne will surely get suspicious. So she wants to get some tests to make sure she doesn't, "have anything, you know what I mean?" If she does, Sheila wants you and Dr. Hernandez to promise not to tell her husband. She asks if there is anything Dr. Hernandez could tell Wayne — if Wayne needs a shot or to take some antibiotics — without revealing that his wife cheated on him?

*This is awful*, you think. *Poor Wayne Brewster's life just took another turn for the worse. His mother may have cancer. His son was in a car wreck and probably has a drinking problem. His boss is a jerk. Now this. No wonder* he *drinks.* You wonder how the Brewster family is able to function at all.

Dr. Hernandez seems equally puzzled. After a few moments of looking at the ceiling, he turns to you. His eyes say it all. *What do we do now?*

TURN TO PAGE 243 TO PLAY ALONG.

TURN TO PAGE 244 TO NOT PLAY ALONG.

# Julie: Play Along

YOU AND HERNANDEZ STEP INTO THE HALLWAY. *Boy, if corridor walls had ears …* you joke to yourself. Hernandez asks your opinion of Sheila's request. You say you believe it might be best, for the time being, to play along with Sheila's charade until the family has had some time to deal with Walter's recovery and Gloria's lesion. Wayne has been under enough stress as it is, you say, and this might just push him over the edge. You suggest a little ruse might give Sheila and Wayne time to work on their marriage, or at the very least, give Wayne time to digest everything on his plate.

Dr. Hernandez agrees with you and gives Sheila a prescription for Wayne. He asks you to set up an appointment for Wayne under the pretense that he needs to come in for some routine blood work to screen for early signs of osteoporosis.

TURN TO PAGE 245 TO CONTINUE.

# Julie: Do Not Play Along

YOU AND HERNANDEZ STEP INTO THE HALLWAY. *Boy, if corridor walls had ears …* you joke to yourself. Hernandez asks your opinion of Sheila's request. You state there is no way either of you should play along with Sheila's ruse. While you know it would be difficult for Wayne to deal with his wife's infidelity while his son is recovering and his mother deals with the possibility of cancer, you also know it would be unethical to lie for Sheila and prescribe medicine for Wayne without him knowing why he was taking it.

You and Hernandez return to the examination room. Dr. Hernandez tells Sheila that he would be happy to refer her to a counselor and can treat her condition, but that he can't treat Wayne without full disclosure. Sheila makes it clear that she is unhappy.

TURN TO PAGE 245 TO CONTINUE.

# Julie: Carla's Visit

IT'S JUST ANOTHER DAY. Before the clinic opens, you and Dr. Hernandez discuss the growing Brewsters saga. You comment on how caring for the family has been challenging. Hernandez remarks that part of any practice involves caring for the "whole" patient. He notes that his family clinic, however, does not include psychological counseling or psychiatry. Therefore, he refers patients to professional counselors when necessary. Dr. Hernandez suggests that the next time a Brewster comes into the office, we consider recommending counseling for the whole family.

While your meeting is in progress, the office door opens and a head pops in. It's Carla, Hernandez' wife, stopping by to "just to say, 'hi.'"

"Hi beautiful." Hernandez is happy to see his wife. "We're just finishing up. Come in."

"Hello, Julie. How are you?"

"I'm well, Mrs. Hernandez. How are you?"

"Wonderful too." She looks around Hernandez' office. "Don't let me bother you two. Just stopped by to drop off some paperwork."

Carla pops out as quickly as she popped in. You sense that she is up to something. You have observed her for three years now and can usually tell when she's up to something.

You and Hernandez finish your conversation about the Brewsters. As you step into the hall, Carla pulls you aside and closes her husband's office door. "Actually, I came by to ask *you* something, Julie, woman-to-woman." Carla is whispering. "I have been thinking about this for a long time now. You know Wayne and Sheila Brewster are close friends of ours … and … I have been thinking that … maybe Wayne has been cheating on Sheila."

You can't believe she is telling you this. *What does she want from me?* You start feeling uncomfortable.

"It's okay, Julie. You can tell me what you know. Enrique and I have already talked about it. Well, he didn't come right out and say it, but I could tell by the way he rolled his eyes when I posed the question to him last night."

You stand is stunned silence.

"Julie, you know this is a close-knit community … and I have suspected that Sheila has been having an affair with her boss for a long time now. She's is always away on 'business,' if you know what I mean." Carla winks. "*Ha!* More like monkey business. Anyway, I can't blame her. With a man like Wayne as her husband, could you?"

"Um, Mrs. Hernandez … I … really can't …"

"I'll bet Wayne cheated on her first, then Sheila found out and decided to get even. *Men!* They're such animals. Anyway, I'm just dying to know who Wayne's mystery woman is? Any idea?"

"No. I mean, I can't …"

"Okay, you're discreet. I understand." Carla's eyes have that wide open, kid-in-a-candy-store look, and it's creeping you out. What's more, you can't believe she is asking you to violate patient confidentiality. You can't even tell her that you have no knowledge of Wayne having an affair because saying a patient said nothing is still a violation of a confidence. One thing for sure, attorney or not, Carla's not one to keep a juicy secret to herself for long. Maybe this bothers you the most. The Brewsters have enough to deal with without having to face a public scandal. Not only will the whole Brewster clan be hurt by a community shunning, but other patients may no longer trust Dr. Hernandez. So many of them are also members of his and his wife's circle of friends. *Oh, bother,* you think. *Hell of a way to start the day.*

TURN TO PAGE 248 TO CONTINUE.

# Julie: The Affair

IT'S MORNING, THE FOLLOWING DAY. You love the way the Austin sun illuminates your kitchen. You make a mental note to make sure you build a house one day with a kitchen facing east. You've just slurped the last bit of foam from your latte, thanking the Lords of Coffee for home espresso machines. *Better than waiting in line for something half as good and five times more expensive.* You allow yourself a brief moment of satisfaction, thinking about all the money and time you're saving by kicking the Starbucks habit, then start digging into your purse. *Where are my keys? I know I ... Oh, there they are.* How is it that keys always fall underneath cell phones? You grab the keys to make a dash for the door when you notice there's a voicemail on your iPhone. It's from Sheila Brewster. *Did she call when I was in the shower?* You hit the PLAY button.

The sound of Sheila's voice is scary. She is screaming. "I've got a good mind to *sue* you, Julie. *SUE THE* HELL *OUT OF YOU.* Do you know what *confidentiality* is? *Do you?*" There's a long pause before Sheila continues. "How *could* you, Julie? *HOW COULD YOU!*" A rude click follows.

*How could I?* How could I *what?* You touch the CALL BACK button. Sheila picks up immediately.

"Hi Mrs. Brewster. I just received —"

"Don't you '*hi*' me, Julie. *You have some nerve!*"

"Please calm down Mrs. Brewster. I don't understand what's going on. I have absolutely no idea what's upsetting —"

Sheila cuts you off again. She explains that she thinks Wayne may have become privy to her affair with her boss. If so, Sheila is absolutely sure that you are the one who told her husband. "It had to be *you*, Julie. Only you and Dr. Hernandez knew, and Dr. Hernandez, *he* would never betray my confidence. *It had to be you*."

You are totally baffled by the assault, which continues. "I'm calling my friend Carla right now, Julie. You know she's an attorney, don't you? Not to mention *your* bosses' *wife*? I suggest you start looking for another job right now, Julie. And for an attorney. You're going to need one."

"Mrs. Brewster, please let me —"

A familiar rude click signals Sheila has hung up.

You stand dazed. You can't believe what you just heard. You want to call Sheila back and figure this out together, if she will let you, but you haven't time. You've got to get to class — right away. Hernandez is out of the office today at a conference and you don't want to disturb him, not until you know more about what's happening. You keep playing Sheila's last tirade repeatedly in your head. *I suggest you start looking for an attorney. You're going to need one.* The bottom of your stomach feels like it's sitting on the floor.

Turn to page 251 to read Clinical Ethics.

# 2: Clinical Ethics

# 2.1: Introduction to Clinical Ethics

MANY ISSUES IN CLINICAL ETHICS are raised in Act 2 of *The Brewsters*. Clinical ethics is less theoretical than philosophical ethics and focuses on issues that arise in the clinician-patient relationship. Where philosophical ethics may spend time discussing pros and cons of theories like utilitarianism and Kantian deontology, clinical ethics tries to clarify concepts like *confidentiality*, *informed consent*, and *patient's rights*.

There is consensus on most of the important clinical ethics issues, and this consensus applies to all clinicians whether in dental hygiene, dentistry, nursing, medicine, or surgery. Indeed, Act 2 centers on clinical ethics to emphasize the value of seeing all clinicians as having similar responsibilities. A single approach makes it easier for people from different professions to communicate with each other and work as a team. Physicians and nurses, for example, collaborate on a daily basis, as do dentists and hygienists. The consensus we present is the result of many sources, including philosophical ethics applied to practical issues, laws (which can vary from state to state), national societies (such as the AMA, ADA, or ANA), and specialty societies within larger professions.

Although the purpose of this section is to help students learn what will be expected of them in clinical practice, we also are interested in introducing new ethical issues. Healthcare is changing in ways that will mean new problems and opportunities. For example, the use of electronic health records (or Electronic Medical Records) in the Affordable Care Act of 2010 and the establishment of a national center to study the comparative effectiveness and the cost-effectiveness of different treatments might be a major boost to the influence of evidence-based medicine. The Patient-centered Outcomes Based Research Institute has been established to speed the clinical adoption of new scientific information. This may challenge some traditional practices. Electronic systems incorporating decision support warnings and alerts may appear to undermine the value of clinical judgment. Patient privacy is also an issue. Some believe patients should be allowed to opt out of having their health information, even if it is de-identified, aggregated into a national database to allow better statistical studies. Act 2 explores the consensus in the field of clinical ethics and introduces some cutting-edge issues.

In *The Brewsters*, Wayne represents a patient's perspective, while Julie represents a clinician's point of view. Since every clinical encounter involves (at least) a patient and a clinician, this helps you see issues from both sides. The following brief instructional summaries introduce many common ethical issues you are likely to confront in clinical settings and clinical training over the next few years. As your training progresses, you will be introduced to the more specialized topics your education requires.

# 2.2: Four *Prima Facie* Principles: Anywhere But New Jersey

IN THE U.S., THE MOST COMMON SYSTEM for organizing ideas in clinical ethics is called *the four principles*: Autonomy, Beneficence, Non-maleficence, and Justice, and have been adapted for medical, nursing, and dental ethics textbooks.

*Autonomy*: Clinicians should provide all relevant information to empower a patient to make an informed decision. Because what counts as well-being is a personal value judgment, patients are the ultimate authority on what is best for them.

*Beneficence*: Clinicians should do what is medically determined to be in the patient's best interest, balancing benefits and burdens. This is a very high standard. It is also altruistic as it rules out letting your own self-interest or third-party interests (e.g., an insurance company) interfere with what is best for the patient. It identifies the clinician as a fiduciary, meaning that the clinician is exclusively devoted to the patient's interest.

*Non-maleficence*, or "First, do no harm." This means the clinician must include preventing or relieving pain and other symptoms in the equation. Quality of life is an important value to protect, not just length of life. This is also a conservative or precautionary principle to avoid heroic interventions that may make things worse. It may counsel that hospice or palliative care is the best of available choices.

*Justice* is the most complex and least intuitive of the four principles. It can be seen as both a positive duty requiring that we give vulnerable people the same care as powerful people or at least try to reduce the health disparities between rich and poor. Justice can also be seen as a negative duty requiring we be careful stewards of resources (so there is enough to take care of everyone). Justice involves each profession's contract with society. This can be seen in a microeconomic way, addressing what we owe to each individual patient (such as a right to basic primary care) or in an macroeconomic way, addressing issues at the political or policy level (e.g. considering what the public as a group owes to all members of the society and, hence, what level of taxes is fair to require everybody to pay).

From an interdisciplinary perspective, it is sometimes useful that four members of a healthcare team maintain balance by each advocating for one of the principles. A doctor might represent Beneficence (the best interest of the patient from a medical point of view), a nurse might be a patient advocate to avoid interventions with high risk or low probabilities of success in the name of "do no harm" (nursing ethics is often called an ethics of caring), and Justice might be the natural domain of the social worker (who often considers financial issues as well as family dynamics and cultural context) or the hospital administration. Decisions should involve the entire team plus the patient, whose Autonomy would represent his or her own preferences in the process.

Even with that interdisciplinary model, it is important that everyone on the team be aware of the importance of all four principles and that no case is only an Autonomy or Non-maleficence issue. The only way to do a good job of understanding is to carefully weigh how all four principles apply. Each principle is considered to be true *prima facie*, meaning always presumed true. Equally important, the four principles are independent, meaning they can conflict with each other. Thus, they are better thought of as helping you understand why a case is complex rather than as a way to simplify a case.

A humorous mnemonic that may help you remember the names of the four principles of Autonomy, Beneficence, Non-maleficence, and Justice is **A**nywhere **B**ut **N**ew **J**ersey.

# 2.3: Ethical Theories & Principles

ETHICAL THEORIES OFFER JUSTIFICATIONS for our intuitions about right and wrong, which help explain these intuitions and predict how we should deal with new problems. The most common theories are Deontology and Utilitarianism, sometimes known by the names of the most famous philosophers who expounded them: Kant (Deontology) and Mill (Utilitarianism). Kant is probably best known for his categorical imperative, which states that you should treat others with respect or as ends in themselves, and not merely as means to your own ends. To put it another way, when a particular action is being considered, you should ask yourself: What if everyone did that? Mill is probably best known for his principle that one should always act for "the greatest happiness for the greatest number of people."

Both theories seek more explanation and justification for ethics than merely citing common sense, intuition, conscience, or a particular religious belief. This is, in part, because there's no way to mediate between two different intuitions except to kick the discussion up to a higher level that allows for comparisons of the two sets of competing personal beliefs.

It is really not necessary for most people to choose between these two theoretical paradigms for ethics, but it is helpful to know they exist, how they differ, and that they both would provide justification for the four principles. The principles are described as "midlevel principles" because they can be justified by either theory. For example, Kant emphasized the importance of Autonomy, meaning (for him) the unique human ability to give reasons for one's actions and to use reason to rise above self-interest. Mill emphasized the importance of liberty, saying that we should all have the right to make decisions (even idiosyncratic decisions) so long as they do not interfere with someone else's freedom.

Principles are, one might say, about halfway in level of abstractness between philosophical theories of ethics and practical issues that arise in day-to-day life and professional practice. Bioethics avoids having to choose one theory or another by using midlevel principles.

# 2.4: Beneficence & Non-maleficence

THE PRINCIPLES OF BENEFICENCE AND NON-MALEFICENCE represent the traditional ethics for the health care professions. That there are two new principles, Autonomy and Justice, in no way means that Beneficence and Non-maleficence are irrelevant. One must still remain true to the altruistic expectations of the profession.

In Research Ethics (in Act 3), Beneficence and Non-maleficence are not distinguished from each other, and only Beneficence is mentioned. You may prefer that approach, only using three principles instead of four. But the guidelines from Non-maleficence cannot be ignored; they are just incorporated elsewhere in that scheme. As with Autonomy, all major philosophical theories of ethics support the importance of Beneficence, even if they provide somewhat different interpretations of and justifications for it.

An important recent concern in healthcare that falls under Non-maleficence is maximizing patient safety and minimizing errors. This includes the use of simple interventions, such as using a checklist to make sure you don't forget any steps in a complex procedure, as well as empowering all members of the team (including medical and dental students, nurses, and hygienists) to prevent anyone from overlooking abnormal findings, correcting dosing errors, or operating on the wrong side of a patient.

Non-maleficence also applies to patient or family requests for inappropriate or futile treatment. Common examples in dentistry might be when a teenage patient says, "Just pull it," and a parent agrees for financial reasons. In private dental practices, which sometimes operate like a cottage industry, it may be tempting to "milk" patients to maximize profit with well-off patients while providing minimal or substandard care for patients with limited resources. Similarly, spending less time on basic restorative dentistry to focus on cosmetic dentistry for its higher profit margin may violate the principle of Non-maleficence.

Offering life-sustaining treatment for dying patients may be yet another example of ignoring Non-maleficence. Any patient who may be in their last year of life ought to have treatment decisions that include symptom management, whether by a palliative care team or the attending. It is also worth pointing out that it is often the nurse who brings this up when it is overlooked, thus highlighting the importance of teamwork and interprofessional ethics.

Non-maleficence is most important in student training situations where one must perform a potentially dangerous or painful procedure for the first time. Here it is ethically incumbent on the trainee to be honest with the patient and to obtain consent. It is equally essential that there be good supervision present in the room.

A final example involves required reporting when one suspects child abuse, spousal abuse, or sexual abuse. Such reporting is expected of doctors, nurses, dentists, and hygienists. All of these are examples of Non-maleficence in action (i.e., trying to prevent harm). In each case, there will be a state agency empowered to investigate, usually a Department of Public Health. Public health is the profession most often in charge of attempts to control social risks such as quarantines for infectious disease outbreaks, whether natural disasters or the result of bioterrorism, but here they attempt to safeguard the health of populations by means of a policy that requires a duty to report. All health professions, in one way or another, articulate and implement their own version of Non-maleficence. Can you think of other examples in your own profession?

# 2.5: Autonomy & Paternalism

WHAT IS PATERNALISM, and when (if ever) is it justified? Has Autonomy gone too far? Paternalism is making a decision for another person with his or her best interests as the goal, but for reasons with which they do not agree (or, sometimes, even know). Its origin is in the parental relationship, and it means treating someone as if he or she were a child. To put it another way, paternalism is treating patients as if they don't know what's good for them, or as if you know better than they do what is good for them. Thus, the motivation behind paternalism is not bad and it cannot be condemned as unethical *per se*.

The reason Autonomy has become predominant among American healthcare professionals is that there is so much variation between people and so much diversity in culture, religion, and values that no one can know a person better than they know themselves. Even if a person makes a so-called "bad decision," or is "wrong" in terms of evidence-based health outcomes, it is often better, ethically speaking, for the person to remain in control of his or her life than to have someone else impose a decision.

What about the feeling or hope that, if you use your authority to force someone to accept treatment, they will thank you in the future? This is an important point. It is very tempting to use this as justification for paternalism. It can be a factor, but it is not enough. That doesn't mean paternalism is never justified, but it takes more than a feeling or hope that a patient will be thankful, because sometimes they are not.

When might paternalism be justified? Balancing benefits and burdens, the benefits would have to be such that no one could doubt them. Burdens would have to be fairly short-lived and mild (such as a week of IV antibiotics), and even that might only be justifiable if the patient could give no reason for refusing (not just one we didn't think was good enough, such as belonging to a religion of which we don't approve) or it seems their decision is based more on indecisiveness and fear than a deeply held belief. Only then might we be able to argue that the patient's refusal is so unreasonable that we would be justified to override it. Clinicians have no right to expect all patients to make what we believe are the most rational decisions.

Do patients have the right to refuse treatment, even when it is clearly beneficial? Yes, but such refusal deserves further discussion to make sure the patient understands everything, particularly the consequences of their choice. After that important assessment, the clinician's burden is not to take their refusal as a personal rejection and to respect the patient's decision.

Here is where some experienced clinicians feel Autonomy has gone too far. They complain that we just offer options to patients and let them make bad decisions, and that represents a lack of caring. They prefer the fiduciary model of the doctor-patient relationship: Physicians and nurses are there to take care of patients and always do what is best for the them. The traditional fiduciary model is ethically very strong, but it is not as contrary to Autonomy as you may think. It is good to care for and about your patients, to put them first, and to make a recommendation, not just to lay out options. Those are all commendable and just what most patients want, but none of those allow a doctor to make the final decision or to force treatment on a patient, especially when it is contrary to what the patient has decided.

Nursing ethics emphasizes the importance of being a patient advocate. Advocacy can require courage, a willingness to tell the truth when others prefer comforting euphemisms, a willingness to listen to patients when they have a sad story to tell, and willingness to speak to a doctor who isn't listening to the patient. Helping the patient make a fully informed decision also can be an act of advocacy.

# 2.6: Confidentiality & Informed Consent

THE IMPORTANCE OF CONFIDENTIALITY is the first lesson taught to entering clinical students. The notion goes back to Hippocrates, a Greek physician born in 460 BC and considered "The Father of Medicine." Hippocrates' students took an oath not to divulge secrets told to them by patients. Clinical fields require practitioners to collect personal information for the benefit of the patient, information they would not provide if they weren't assured it would be kept confidential. To violate that tacit promise would harm not only that patient, but potentially undermine trust in the profession.

After an introduction to confidentiality, the second ethics topic taught is often informed consent. While confidentiality is an ancient doctrine, informed consent is considerably more a modern idea. It might be thought of as the legal support for the ethical principle of Autonomy.

Informed consent started in 1914 when an influential judge named Benjamin Cardozo said in a court case called Schloendorff, "Every human being of adult years and sound mind has a right to determine what shall be done with his own body." [13] Another legal opinion, called Salgo, coined the phrased "informed consent" in 1957 and said that physicians have a duty to disclose the risks of treatment to a patient, and that a patient must give permission before a proposed treatment occurs. [14] This was in accordance with views enunciated by the Nuremberg Code after the 1949 trial of

Nazi doctors and nurses. The Nuremberg Code began by asserting "the voluntary consent of the subjects is absolutely essential."[15] Salgo changed the nature of the wrong or harm caused from being a case of battery (touching someone without their consent) to "liability if he withholds any facts which are necessary to form the basis of an intelligent consent by the patient."

What does informed consent require today? The best advice is to tell patients that full and fair disclosure is ethically and legally required. For the most part, if you would want to know something before deciding, then you should assume patients would as well. First, include all the reasonable treatment options, not just the one you think best (this is still restricted to the reasonable options, not unproven or inappropriate options). Most patients also will appreciate a recommendation. Second, you must include the option of choosing not to treat the condition at all (the right to refuse treatment). Third, include the most likely benefits and burdens (risks and benefits) for each of the reasonable options, including things like the length of rehabilitation and the expected rehab site. Last, but not least, the clinician must make sure the information is understood by the patient. This final element raises the issue of a clinician's communication skills, which are important to refine during your years in school.

While there are some possible exceptions to informed consent, they are rare. For example, there might be people from other cultures who have never been allowed to make their own decisions and aren't comfortable with the idea. They can defer their consent to someone else, but they have to be the one who makes that decision, not their family and not their doctor.

It is important to realize that informed consent is a process of communication. Its purpose is to help patients make informed decisions and to protect patients from coercion. Informed consent is not a piece of paper to get signed or legal protection for the healthcare team. Informed consent isn't like being a salesperson or parent, but more like an educator and advocate. Informed consent requires a willingness to tell the truth and not hide behind comforting euphemisms or technical jargon. Its original intent was to encourage clinicians to be patient centered.

# 2.7: The Right to Refuse Treatment

HERE'S ONE OF THE FIRST LESSONS that follows from the recognition of Autonomy: Patients with *decision-making capacity* (DMC) have the right to refuse treatment. This is true for any treatment, including life-sustaining dialysis, use of a ventilator or feeding tube. The clinician's job is to make certain the patient understands the situation and tries to persuade the patient to make what the clinician thinks is the wise choice. After that, it is the patient's decision whether to accept the recommendation.

Nephrology was the first field to discuss letting patients refuse life-sustaining treatment after dialysis became widely available in the 1960s. It is the only treatment for which the U.S. government will pay for any citizen. Once the cost was realized, however, the U.S. government never made that guarantee for any other treatment. This universal coverage meant that many people had a chance to live on dialysis long enough to realize that life on dialysis involved many compromises. Now, 10 to 15% of all dialysis patients eventually decide to stop treatment. [16] It is usually a fairly quick and painless death. The patient enters a coma in about a week and dies within two weeks. Most suffering can be avoided just by avoiding fluid intake.

Much of the literature on refusing life-sustaining treatment is based on Jehovah's Witnesses' right to refuse blood transfusions. So long as they have Autonomy (DMC), they can do so, even if others think it is unwise. It is important to understand that the right to refuse treatment does not hinge on it being based on a religious belief. Atheists and agnostics have the same rights. It is also important to understand this is the right to refuse treatment for oneself. It does not imply that adults can refuse life-sustaining treatment for their children. In the words of the Court (Benjamin Cardozo), "Parents may be free to become martyrs themselves. But it does not follow they are free, in identical circumstances, to make martyrs of their children ..."[17]

The right to refuse treatment applies to all fields. Patients must give consent before you begin treatment, and they always have the right to withdraw consent later. If that were not the case, they might be much less willing to consent ("give it a try") in the first place, leading to the unintended consequence of fewer people accepting an indicated and potentially valuable treatment.

# 2.8: Rights of Children & Teens: The Capacity to Make Decisions

IN *THE BREWSTERS*, WE SEE A MOTHER, Sheila, bringing her daughter Stephanie to a physician for birth control pills, even though it isn't clear her daughter wants them. The most important person in this scenario is the patient, Stephanie. If Stephanie is to make an informed decision, then Stephanie should know all the effects of birth control pills, including the risks. She should know, for example, that she should not smoke if on "the pill."

At what age should you think of patients as having a right to a say in their health care? While you can't get consent from babies or young children, you should share information with them starting at around age 6 or 7, and try to get their agreement (called "assent"). Physicians and nurses in pediatrics and family medicine usually begin seeing children for part of the interview without a parent in the room starting as early as age 6 or 7 to start normalizing the right to privacy and confidentiality for both child and parents. This starts building a child's trust in the clinician. It also helps the child know there is an adult concerned about their welfare along with their parents.

At what age might a minor be able to consent to medical treatment? If they are a teenager, then clinicians should be thinking about the possibility. Alternatives can be worse (i.e., kids likely to run away from home). Regarding terminally ill children, the American Academy of Pediatrics says, "probably most minors by the age of 14 have the capacity to make decisions about life-sustaining treatment."[18]

A few facts and lessons about birth control, abortion, and sexual health are worth pointing out. First, 18 is the age of legal adulthood (age of majority, no longer a minor) in most states. However states have an exception to the age regarding birth control and treatment of STIs and STDs. This can be justified either by the *mature minor doctrine*, which recognizes that any minor capable of informed consent has the capacity to make their own decisions, or by a simple modification of the law that exists in most states, including Texas.

Second, considering parental notification laws for abortions for minors, it is important to help minors learn about birth control and sexual health. It is also important to know that: (1) there is no parental notification law for birth control; and (2) every state must have an option of judicial bypass for cases where a minor fears the consequences of parents finding out she is considering ending her pregnancy.

Third, a critical lesson for clinicians morally opposed to abortion and not able to counsel patients fairly on all medical options is referring patients to a reliable and objective agency. This professional responsibility is endorsed by many professional organizations, such as the International Federation of Gynecology and Obstetrics.[19]

DMC is best judged the same way for people of all ages. Do they understand the diagnosis and the reasonable treatment options? Can they make a choice and understand its consequences? If a patient is capable of giving informed consent, then the patient has DMC. Still, this does not mean you should disregard a teenager's parents. Rather, you should encourage parents to see maturation of their child as a good thing, not a challenge to their authority, and to give teens increasing responsibility over their life, including medical decisions, as they mature.

# 2.9: Telling the Truth: Breaking Bad News

ONCE UPON A TIME, it was common to keep bad news from patients. Practices changed quickly during the 1970s, perhaps in step with general changes in American culture then, including the development of bioethics. Studies show that in the early to mid 1960s 80 to 90% of patients were not told their cancer diagnosis. By the mid to late 1970s 80 to 90% of patients *were* told their cancer diagnosis. That's a dramatic change in one decade.[20]

Why was the truth kept from patients? The reason often given was that it would upset the patient too much and might even lead to their giving up hope. The right to keep the truth from patients was based on the clinical judgment of the patient's best interest, called *therapeutic privilege*. However, when asked, most patients said that they would prefer to know — after all, it wasn't as if the bad news would go away if they didn't know.[21] Knowing would at least allow them to make the plans for the remainder of their life: what to do, where to go, whom to talk to, and what to say. Now, even though therapeutic privilege is still quoted in legal doctrine, it is rarely invoked.

Today, one may hear arguments for keeping a diagnosis hidden from a patient on the grounds of being sensitive to cultural differences. This is sometimes claimed, for example, in reference to Latino, Italian, Eastern-European, Southern European, and Asian cultures. If there's any truth to these claims, it depends on how long the patient has been in the U.S. and acculturated to our social norms. If he or she came to the U.S. after reaching adulthood, speak poor English, and are very traditional, then one may wish to approach things this way: Tell the patient you have important information about their health and ask if they would like you to share it for decision-making purposes; or ask if there is an alternate decision maker whom you should tell instead. For most people, however, the avoidance of truthful and meaningful discussion only increases anxiety and is harmful. Discussion is often therapeutic. At the very least, never presume to know what patients would like to know without asking them.

# 2.10: Telling the Truth: Work Excuses & Disability

SHOULD CLINICIANS BE SYMPATHETIC TO PATIENTS working in hostile work environments with unfair personnel policies? Should those sympathies extend to the point of lying for the patient, so long as it is a "white lie"?

Being asked to give a patient an excuse for something like missing work or not taking a flight is the kind of everyday situation that clinicians dislike, but often don't recognize as an ethical dilemma. The conflicting arguments about what is best to do (the definition of "dilemma") are to help your patient and to tell the truth.

When a patient requests a health professional to fill out a job or government form (work injury being the most common), it is easy to get annoyed. Before losing your sympathetic fibers (an ethics pun), remember that even if this is the fifth form request of the week, it is the first from this patient. A "GI bug" or types of pain (back pain, migraine, etc.) which cannot be objectively verified are often the cause of work absences. After hearing them often enough, one can build up a suspicion that they are all really just excuses to avoid work, especially since jobs most likely to create physical pain also are generally the least enjoyable. Still, since there is no way to know the subjective experiences of another person, you must always begin with the charitable assumption that they are not lying.

Sometimes the employer seems so cruel that you want to take the patient's side even if it means lying. Patients will sometimes ask for "little white lies," and they can seem to help patients deal with unfair rules. It is rarely a good idea. A clinician's job cannot be to undo all the injustice in the world. When you find something worth fighting over, it is usually better to confront the issue honestly rather than lie to get a person out of trouble.

When it comes to work related injuries, this dual role (patient advocate and truth-telling clinician) can be very uncomfortable. Complicating matters is that it is not always possible to know how a patient was first injured (for example, back problems). There will be clinicians hired by the company to investigate a claim, and they will be skeptical (and paid to disprove the claim). This makes the process of investigating work related injuries an adversarial system in which you are presumed to be an advocate for the patient. There are two principles worth following: (1) never knowingly lie for your patient; and (2) try to be a good advocate for your patient as far as your knowledge allows.

Are there ways to do that? For a patient looking for a note justifying a work absence, you might say, "I don't feel comfortable telling your employer a lie and I don't think you really want a doctor who would do that." Then, write something like:

> To whom it may concern:
>
> Mr. John Smith was seen and evaluated in my office today. He may return to work without restriction.
>
> M. Brown, MD

Notes like this nearly always suffice for an employer.

Some students worry about whether they can get into legal trouble for a white lie. This should not be your primary concern. It's a sign that you might make medical decisions based solely on liability issues. "Defensive medicine" is practicing in ways that reduce exposure to potential lawsuits, such as avoiding high-risk patients and ordering extra tests. Defensive medicine increases health care costs in ways not necessarily beneficial to patients. Defensive medicine is bad medicine.

# 2.11: Telling the Truth: Admitting Mistakes

THERE HAS BEEN A MAJOR MOVEMENT in hospitals to tell patients the truth when a mistake occurs, largely to find better ways of identifying mistakes and preventing future ones. The effort has been enhanced by the establishment of a hospital administration field known as Quality Assurance (QA) or Continuous Quality Improvement (CQI). The field focuses on how to improve patient safety, especially since 1999 when the Institute of Medicine Report, "To Err is Human," estimated there are 400,000 preventable injuries and deaths in U.S. hospitals every year, and even more in nursing homes. [22]

In some errors, there is no harm to the patient. This is sometimes called a "near miss" in the hospital. When there is no harm, should we bother to tell the patient? While some have argued there was no need to inform patients in such cases (and that it might cause them to worry unnecessarily), this is no longer considered acceptable. [23] If the medical or dental culture were to allow clinicians to not admit to their mistakes, the mistakes would be more likely to happen again.

It is best to be honest. It used to be that hospitals were afraid of being sued if they admitted to mistakes and their legal advisors or Risk Management sometimes even counseled that little good could come of "too much honesty." The legal tide probably turned when research showed patients were less likely to sue when told the truth about a mistake accompanied by a sincere apology than if they discovered the mistake on their own. It is the suspicion of a coverup that incenses patients and provokes them to exact Justice (or revenge). If a doctor or dentist shows remorse, patients will sympathize with them and, usually, forgive them. Clinicians who are most likely to get sued, in contrast, are those who appear arrogant, busy, and unsympathetic.[24] Nurses can help the healthcare team by advocating for more honesty with patients. The same applies to hygienists who can talk to dentists about the value of honest disclosure.

Expert clinical judgment can address how to deal with stresses caused by an exaggerated fear of lawsuits. Here are a few suggestions: First, realize that you only make things worse by worrying about it all the time. Second, remember that the best defense in court is not, "I did it because I was trying to avoid a lawsuit," but, rather, "I did it because it is what I thought was best for my patient." Third, it may be best to simply assume that you will probably be sued sometime in your 40-year career (or at least threatened by a suit), so don't sweat it. When the time comes, get a good lawyer and do not let the experience ruin your mental health or your love of the job. Lastly, if you do make a mistake, admit it and apologize. It will make you feel better from the sense of no longer having something to hide, and may even prevent you from being sued.

# 2.12: COI: Evidence-based Medicine & Cost Effectiveness

BORN IN 1849, CANADIAN PHYSICIAN SIR WILLIAM OSLER is regarded as "The Father of Modern Medicine." A founder of Johns Hopkins Hospital and medical school, Dr. Osler established the first physician residency program for specialty training. He is quoted as saying, "If you listen to your patient, they will tell you their diagnosis."

In the midst of growing technological prowess, humility remains an important virtue for physicians, just as it was a century ago in the time of the great physician, Sir William Osler. What appear to be a patient's exaggerated worries should never be dismissed. Sometimes, exaggerated worries turn out to be right. Asking a few good questions and giving reassurance that the patient's symptoms don't correlate to the disease they fear should reduce their anxiety. Most patients have faith in their doctors, dentists, and nurses. Note the term "faith" here. This is part of a clinician's therapeutic armamentarium not based on pharmacologic or invasive intervention. Sometimes it's considered a placebo effect, other times the "priestly" power of clinicians. If reassurance doesn't suffice, it still can be acceptable to do some blood work or other low-cost tests if necessary to put a patient's mind at ease.

That said, expensive tests, screening tests for very rare conditions, and risky and potentially harmful tests are another story. The principle of Non-maleficence is the oldest and most inviolable in clinical ethics. Doing something that makes things worse should be avoided at all costs for ethical reasons. Similarly, driving up costs is irresponsible and can lead to serious legal consequences, such as penalties from practice plans or fraud charges from the government.

*Evidence-based medicine* uses the best current scientific evidence in making clinical decisions about patient care. Factors include reliability and cost (both direct and indirect).

What if the patient is rich and offers to pay out of pocket? If you allow that to happen too easily or too often, it promotes the impression that your professional opinion is not very reliable. It also reduces medicine and dentistry to something less than professions based on a fiduciary responsibility to patients. Instead, they become commercial transactions. The presumption should always be: Do nothing that is not indicated or reasonable. Practice guidelines are there for good reason (i.e., scientifically validated and evidence-based) and should be followed for all patients.

In dentistry as well as in medicine, one can be overly aggressive with testing (like X-rays, cone beam CT scans) or treatment, sometimes creating iatrogenic (illness caused by treatment) side effects. The profit motive has a way of distorting everybody's judgment. Studies have shown clinicians with labs in their office do more tests than clinicians who don't have labs.[25] The same has been shown for imaging equipment, including DEXA scans.[26,27] It is likely that enhanced profits are often part of the motivation for testing. Once you make an investment, you want, or even need, it to pay for itself. Therefore, you must be very careful. If a patient asks for a test you consider unnecessary, remember that it may sometimes be justified for the patient's sake, but never solely for the extra income it might generate. The latter is a conflict of interest. In the story, Wayne Brewster is portrayed as a possible hypochondriac when his request for a DEXA scan is not indicated. That the doctor owns the DEXA scan equipment is a temptation. If the DEXA scan is justifiable at all, it is because it is low-cost, low-risk, and might put Wayne's mind to rest, at least temporarily.

Every clinician has legitimate self interests that sometimes conflict with those of their patient. As a health professional, you are supposed to put your patient's interests ahead of your own, which is not always easy to do. The goal might best be described as managing conflicts of interest by avoiding situations where the temptation is to put self interest first and being honest about conflicts when they exist.

# 2.13: COI: Cultural Diversity & Accepting Gifts From Patients

THERE ARE MANY CULTURES in which it is customary for patients to give gifts to clinicians. To refuse a gift in some cultures is dishonorable. In the U.S., however, accepting gifts from patients is discouraged.

What if it's a small gift? Shouldn't one just take it? After all, doesn't this build rapport? No, there are problems with accepting even a small gift. First, you won't know if it's a small gift until you open it. How would it look if one opens, inspects, and then rejects a gift? Second, what one person considers a "small gift," another may consider large. You won't know if your patient expects some sort of special treatment as a result. Any patient who expects special treatment (including friends and family members) can lead to ethical complications later. Third, there is the problem of whether accepting a gift might influence your behavior. It is common to assume gifts won't influence us, but studies have shown people respond to gifts by feeling grateful and indebted to the giver.[28,29,30] They subsequently look for a way to repay the gift to show appreciation. This would violate the fundamental principle that clinicians should be nonjudgmental and treat all patients the same.

Does this mean you never accept any gifts? Exceptions should be rare. It would not be wise to accept a store-bought gift from an American patient since American culture does not have a custom linking honor or dignity to gift giving.

Exceptions might be something homemade or handmade, such as cookies or a special dish from your patient's culture. Any refusal of gifts must be made with carefully chosen words that express your feeling of gratitude for the offer as well as your respect for the patient.

# 2.14: COI: Gifts from Companies

MANY DRUG AND MEDICAL DEVICE COMPANY marketing practices have come under ethical scrutiny. Questions have been raised about providing clinicians and researchers samples, gifts, and sponsoring symposia, research and journal articles. Pharmaceuticals marketing directly to patients (spending almost $5 billion in 2007 for *direct-to-consumer*, or DTC advertising) has also been scrutinized.[31]

It is important to be aware that profit motives can be a powerful counterforce to the evidence-based practice expected of clinicians and researchers. Drug and medical device company representatives are often attractive young individuals recruited from communication and business schools. They try to persuade clinicians by various means (e.g., offering free lunches and pens, paying for textbooks and courses, etc.) and presenting well-honed sales pitches practiced through role playing. Their objective is clinicians and researchers using their company's product. New products may be more expensive but less well studied than older ones. Companies are in business to sell products, especially profitable ones still under patent protection. Some Phase IV trials may appear little more than marketing schemes that pay physicians to switch patients to a new drug with a promise of a follow-up survey to be sent later.[32]

An important issue is company-provided free samples. This practice is sometimes defended because clinicians could, theoretically, give them to the poor. However, studies show many free samples are used by the clinician or clinician's family.[33,34,35] That makes free samples seem more like bribery than charity. However, what if you, indeed, had a sizable number of poor patients and gave all your free samples to them? Does that make it justifiable? Perhaps, but remember that free samples are usually a "starter pack." What happens when the supply runs out?

Another issue involves journal articles which pharmaceutical representatives sometimes provide clinicians as marketing materials. Knowing that clinicians are often short of time, company writers may produce abstracts worded to make weak results sound promising. It is best to read (at a minimum) the beginning and end of any article, not just a synopsis. Also, some presentations by clinicians or researchers at company-sponsored scientific meetings are prepared by the company for the expert, yet the company may pay the expert an honorarium. There have been dozens of scientific journal articles that were withdrawn when it was discovered their authors had been paid by pharmaceutical or medical device companies. Other articles have been quashed by companies paying for research whose outcomes do not back up company claims. Professional journals often have writing guidelines to help authors minimize distortions that could influence clinical judgment.

The lesson here is that free trips, free meals, free pens and pencils, and free samples aren't really free. They are paid out of company profits, which means someone ultimately pays for them, like insurance companies, taxes, and patients. They are part of the reason U.S. health care costs exceed almost every other country relative to gross domestic product.[36] As the saying goes, "There's no such thing as a free lunch."

# 2.15: Complementary & Alternative Medicine

HOW SHOULD CLINICIANS RESPOND to the growing use of Complementary and Alternative Medicine (CAM) by patients? Must patients provide scientifically valid and evidence-based reasons for their decisions? How should we respond to requests based on religious beliefs?

Common CAM treatments include massage, prayer, hypnosis, yoga, and meditation. Related to CAM is the organic food movement, which goes back to the 1940s and really came to prominence in the 1960s. Interestingly, decades of research on nutrition have proven much of the claims of these movements correct. For example, the FDA now recommends more whole grains in our diet. Perhaps the same will be shown with other CAM treatments.

CAM treatments mentioned here are unlikely to have direct toxic effects, but sometimes they can have indirect toxic effects — they may delay a proven treatment or have unknown drug interactions. Thus, it is important not to dismiss or to belittle CAM so that your patients will be comfortable talking and sharing with you. Clinicians should be willing to allow simultaneous alternative and complementary treatments so long as they don't interfere with medical treatment. These can include support groups, prayer, relaxation, meditation, and herbal remedies. Demonstrating understanding of your patient's needs can play a significant role in the patient trusting you.

It's sometimes useful for the clinician to provide a patient a list of questions to ask their CAM provider: Has the provider treated patients with this condition/diagnosis before? Do they know of side-effects or dangerous interactions with other medications? How many weeks should it take to see some benefit? Is the CAM provider willing to give other health professionals updates? These kinds of questions can facilitate useful dialogue.

When should CAM be challenged? When patients or CAM providers want to substitute unproven and scientifically baseless therapy for a proven effective treatment, especially when a condition is life-threatening. In particular, you must draw a line between adults and minors, intervening to protect children from poor decisions with potentially serious and permanent consequences.

Conversations about CAM can be difficult to negotiate. The goal is preventing health care professionals from becoming so frustrated that they "flee to medicine-land"[37] and keeping patients from "fleeing from it." Good clinicians do not drive patients away or drive their practices underground.

# 2.16: Justice: Preventive Medicine & Vaccines

VACCINOLOGY IS ONE OF THE GREAT FIELDS of biomedical science. It crosses the disciplines of basic science, public health, primary care, and public health nursing. Vaccines probably save more lives than all the Emergency Rooms in the country.[38] Indeed, the only intervention that has saved more lives is water chlorination, another public health and preventive medicine intervention.[38] The most successful outcomes in dentistry are thanks to fluoridation of water.[39] Yet there are scares among the general populace about the safety of vaccinations, as there was about fluoridation.

Fear of vaccines was not allayed when British physician and researcher Dr. Andrew Wakefield argued for a connection between vaccines and higher rates of autism. There were allegations of financial conflicts of interest in his research.[40] The British General Medical Council investigated and ruled there was dishonesty in his publications.[41] Dr. Wakefield has been stripped of his license to practice in the U.K.[42]

Why do people believe sensational claims about vaccines and autism? The best explanation may be that people see causation when two events occur around the same time, even though the events could be unrelated. For example, the first symptoms of many diseases are observed around age 6, but that doesn't mean vaccines required before starting school (around age 5 or 6) have anything to do with these diseases.

Another explanation is that statistical reasoning is hard for some people to understand. If one begins with harmless sounding beliefs, such as "everything happens for a reason" or "every effect has a cause," then: (1) you are looking for causes, real or imagined; and (2) you are likely to find statistical explanations of "random events" unsatisfying.

There is a shorter average lifespan in the U.S. than many other countries, including many less affluent countries.[43] The fear of vaccines sadly contributes to many avoidable deaths of children in the U.S.[44] Within the U.S., some states do better than others in preventive health services. Other causes of morbidity include smoking, alcohol abuse, impaired driving, gun violence, drivers and passengers in cars without seat belts, and riding motorcycles without helmets. Assertions of individual rights to be free from legal mandates add up. What is most unjust is when innocent bystanders suffer as a result, whether due to secondhand smoke, a pedestrian hit by a drunk driver or a child infected by a disease because of scientifically unfounded fears.

# 2.17: Justice: Access to Care & the Role of Government

ACCORDING TO THE 2010 U.S. CENSUS, 50 million people in the United States have no health insurance.[45] Medicaid is a government insurance plan for the people with low income and their children. Requirements vary from state to state, with wealthier states typically paying to let more people qualify (e.g. up to 150% of the federal poverty level), while other states (including Texas) require being below the federal poverty level. What is covered also varies, with some states including more generous options. Medicaid also covers dental care for children and administers the state children's health insurance program (sCHIP). Money comes from federal and state taxes. The Affordable Care Act of 2009 proposes to subsidize Medicaid through higher incomes to insure millions more people.

If you do clinical rotations in a private practice, you may find doctors or dentists who do not take Medicaid. Many professional schools find it hard to get enough preceptors, and so do not require clinical faculty to see Medicaid patients. In 2006 only 38% of Texas physicians accepted all new Medicaid patients.[46] Each time state budgets force cuts in Medicaid payments, the percentage of doctors and dentists accepting Medicaid patients declines.

Medicare is different from Medicaid. Medicare is primarily for people over 65, and for disabled people or those needing dialysis. Medicaid, Medicare and the Veterans Administration (VA), provide insurance for tens of millions of Americans.

Federal and state governments now cover around 50% of the cost of care in the U.S.[47]

Even Medicare, which pays better than Medicaid, is only accepted by 60% of doctors in Texas.[48] Medicare is often viewed as a benchmark for reimbursement rates by private insurers. That means that Medicaid reimbursement rates are lower than most private insurance plans. This is often mentioned as a reason for not taking Medicaid, as well as a paperwork burden. There may also be other factors, such as not wanting one's waiting room to be filled with "poor people."

Unfortunately, the unspoken message that many students take away from their clinical experience is, "Non-sponsored" (uninsured or no pay) and lower income (under insured) patients are seen in free clinics and public county hospitals, while wealthy and insured patients are seen in private practices. The unintended message might be, "Better insured patients get better treatment." This implies what many consider an ethically unacceptable two-tier system, allowing socioeconomic biases to have an effect on quality of care. Clinicians should treat patients at public and private hospitals equally.

There is a very long history of clinicians regarding their ethical duty as taking care of all patients, regardless of ability to pay. The AMA, for example, states that physicians have an obligation to provide services to the poor and the uninsured until there is some form of universal health insurance.[49] If you become a practitioner, consider setting aside some time every week or every month to provide care for the uninsured and under insured. You can do it in your office or in a free clinic like neighborhood health services. A reasonable goal might be giving 10% of your time at no charge. That would be one afternoon per week, for example, or one day every other week. Think of it this way: Many mosques, churches, and synagogues recommend *tithing*, the practice of giving a percentage of one's income to charity. Given your education and training, the most valuable commodity you can give is your time and skill. You might find your self-esteem gains greatly from this small dilution of income.

Assuming we would not force physicians and dentists to accept uninsured or under insured patients, or to live and practice in rural areas, how can the problem of access to care be addressed? One idea is adding Texas to the list of states that allow advanced practice nurses or Nurse Practitioners to diagnose and to prescribe without the oversight of doctors.[50] Under the "patient-centered" model being encouraged by health reform, primary care providers, including nurses, can lower healthcare costs by increasing preventive care.[51] A 2010 Institute of Medicine report, titled "The Future of Nursing: Leading Change, Advancing Health," urged state officials to expand what nurse practitioners can treat.[50] The report found health systems that increased nurses' responsibilities delivered "safe, high-quality primary care."

This could really help Texas, since 171 of Texas' 254 counties are designated wholly or partially medically underserved.[52] Texas has more than 12,000 advanced practice nurses, including 7,900 nurse practitioners who can help with the access problem. If the idea works, and if the number of dentists accepting Medicaid remains low, some might wonder if we also should consider a program of advanced practice dental therapists to do routine exams and fillings?[53,54,55]

# 2.18: Justice: Health Disparities & Public Policy

PUBLIC HEALTH RESEARCH has revealed some surprising results on how socioeconomic conditions influence health. Disadvantaged groups, such as racial minorities and the poor, receive inadequate healthcare for many reasons, including lack of insurance, lack of doctors who take their insurance (Medicaid), cultural and language barriers, and difficulty getting to appointments by means of public transportation.

A commonly cited article claims 70% to 90% of illnesses is preventable.[56] Preventable factors may be due to air and water quality, and lifestyle choices like smoking, dietary, and exercise habits. Some of these have statistical correlations to socioeconomic class. The result is that income and education are better predictors of life expectancy than medical insurance coverage or quality of healthcare. These important factors are addressed more by Public Health than any other health profession, and they have great clinical significance.

Public Health research also reveals that greater income disparity between the rich and the poor in a society has negative consequences on the health of populations.[57,58] Much of this is thought to be due to stress. Everyone is better off in a society with a large middle-class, people with self esteem, and no group feeling like second-class citizens. These data support thinking of the principle of Justice more broadly. To improve human health, we need to think beyond distributive justice (for example, equal access to care) to social justice. Good education and social policies that provide people with job and financial security also matter to health. To pursue these goals, it is important to control the costs of healthcare.

The U.S. has always had a self image as a place where people can climb social and income ladders, a land of opportunity. Many ethicists as well as economists hold that healthcare should be seen like public education. We need to offer a decent level of education and healthcare to everyone, enough to allow people a fair opportunity to improve their station in life. If the poor are given the worst schools and the least healthcare, we risk becoming a rigid class society rather than one where upward mobility is a possibility.

Most of these issues are studied in Public Policy courses. Clinicians often will pursue a Master of Public Health (MPH) degree to complete their education concerning justice and healthcare. On how poor health corresponds to social factors like income inequality, there are two authoritative websites, the World Health Organization[59] and the U.S. Centers for Disease Control and Prevention.[60] You may also wish to consult the Institute of Medicine (IOM) of the National Academy of Sciences report on health disparities.[61]

TURN TO PAGE 305 TO READ ACT 3.

# ACT 3

"What shall it profit a man if he gains the whole world
but loses his soul?"

*~ Jesus*

# Who Am I?

YOU'RE SURE GLAD THAT YOU ARE NOT WAYNE BREWSTER. Wayne's rebellious son, Walter, started medical school this year but things aren't going well. Walter wrecked his car leaving a professor's reception for incoming medical students — drunk. A few weeks later, Wayne himself was ticketed for DWI, a result of getting into trouble at work for unexcused absences. As a mail carrier, Wayne's losing his driver's license would be synonymous with early retirement.

Wayne is beginning to think his troubles are his wife's doing. Sheila was never much of a stay-at-home mom — more like a stay-*away*-from-home wife, the kind always on the go, flying somewhere on business. Now, Wayne knows why. Sheila's been having an affair with her boss, "Armand." Nobody believes that's his real name. Too Hollywood. Lord knows how long *that's* been going on. To cap things off, Wayne's mother, Gloria, has developed a suspicious lesion on her tongue. Wayne got an appointment for his mom with a specialist to check it out, but Wayne is convinced it's cancer. He is equally sure his mother will die of worry before the cancer gets her. Then, there's Wayne's nagging fear he's developing osteoporosis …

Yeah, you are sure glad you are not Wayne Brewster. In fact, you would much rather be …

TURN TO PAGE 308 TO BECOME WAYNE'S MOTHER, GLORIA BREWSTER.
TURN TO PAGE 381 TO BECOME PARVESH SINGH, DENTAL STUDENT.

# Gloria: The Appointment

YOU AND YOUR SON, WAYNE, arrive at the Dental clinic. You are glad that Dr. Hernandez has referred you to Dr. Nguyen. You have always trusted Dr. Hernandez' judgment. If he thinks a dentist should look at that little white spot on your tongue, than so be it.

When you arrive, you are greeted by Leticia Roberts, a pleasant young woman who introduces herself as Dr. Nguyen's dental hygienist. "We have some forms for you to fill out." She hands you a clipboard but Wayne grabs it out of your hands. You stand at the check-in counter as Wayne quickly completes the paperwork. As he works, you notice a business card on the counter. Picking it up, you see that Dr. Nguyen is an adjunct faculty at the dental school. *That's reassuring,* you think.

Wayne hands the clipboard back to Leticia.

"Done so soon? Very good Mr. Brewster. Okay, Mrs. Brewster. Please come right this way."

Leticia leads you down a hall. Wayne follows. You take a seat in an examination chair and look back at Wayne. *Poor guy.* He's nervous and tired. You feel sorry for him. *That divorce! It's gotta be tearing him up. Heck, it's tearing me up, Ugh. What the kids must be going through* … It tears your heart out. Now Wayne has to worry about you, too.

You open your mouth as Leticia adjusts your chair. She positions the light. First here. Then there. Then back again. She pokes a little mirror around your cheeks. As she examines your mouth, she records her notes. "Well, there are no cavities." Leticia withdraws the mirror.

"I would hope not. I've got dentures."

"I know. It's just a joke."

You decide you like her. "So, what is it?"

"Dr. Nguyen will be right in. She'll give you more information."

You feel as if she's not telling you something. There's a little pang in your stomach, and for the first time, you get a little worried. *Could this be serious?*

Leticia excuses herself. Within a few moments she returns with Dr. Nguyen.

"Hello Mrs. Brewster." You are greeted by a petite, vibrant, Asian woman with a forceful handshake. "Okay, let's see what's going on in there." She examines your mouth as Leticia did just a few minutes ago. Except, Dr. Nguyen is spending more time probing around your gums and cheeks. "Well, … I do see and feel something … just not sure I know exactly what it is."

Wayne nervously asks, "What do you *think* it is?"

"There are several possibilities, one of which is nothing."

"Verascope?" Leticia asks Dr. Nguyen.

"Not sure it will tell us much, but it couldn't hurt, I guess."

"What's that?" Wayne asks?

"A simple test that looks at the cells in the mouth," Nguyen explains. "It will just take a couple of minutes."

"Well, I guess I got a couple of minutes," you joke. "Just don't make me late for the rodeo. I'm in the afternoon calf-roping event." Everybody laughs.

Leticia puts your chair in a reclining position. She unsnaps a small handheld device from a box and hands it to Dr. Nguyen. Leticia then gives you a pair of big sun glasses to wear.

"Here we go." Dr. Nguyen holds your jaw open with one hand and shoots a blue light into your mouth. Nguyen looks through one end of the device as she tours your mouth. "If anything bothers you, let me know."

You can't help but be amazed by modern technology. You used to think that remote controls were ingenious, but this ... *holy moly!*

In no time at all, Dr. Nguyen is finished. "Hmm. Not sure we know any more than we did a few minutes ago. Somewhat inconclusive." She turns off the device. "I still think it is nothing, but I want to be sure. I'm going to refer you to the Crenshaw Center for some tests. Leticia can set everything up for you, if that's okay."

"Sure," you agree. "But — and I'm being serious now — try to make it in the morning. I play Uno in the afternoons. I don't mind missing my aquatics, but I don't want to miss my card game if I can help it."

You and Wayne adjourn to the lobby as Leticia makes a phone call. You wish everything had been resolved today, but what can you do? Nada. You'll have to wait. So will Wayne. That's the hard part.

Leticia returns with a slip of paper. "Here's your appointment for Tuesday. It's first thing in the morning. Have a great rest of the day."

As you gather your things, you perk up. You still have enough time to get to your card game this afternoon. "Hurry up, son. The girls are waiting. We can drive through Whataburger on the way home. My treat."

TURN TO PAGE 312 TO CONTINUE.

# Gloria: The Exam

*WELL, IT'S BEEN A GOOD LIFE.* You can't get that thought out of your mind as you and your son, Wayne, arrive at the Crenshaw Center. While you hope for the best, you can't help but worry that this might be *it*. You're sure Wayne thinks it's cancer. But if you do have cancer, and if there is nothing they can do about it, well then, *so it goes*. You're prepared for death. You've lived an honest life. You've raised a little hell and a good son — a bit on the neurotic side — but nonetheless, a good man. You've enjoyed your grandkids, especially the oldest, Walter. You are so proud of him. You had a successful career as a painter when you were young and made many wonderful friendships along the way. You were hoping to hang around to see some great-grandchildren, but if it ain't in the cards, then *c'est la vie*.

"I signed us in, Mom. They said it wouldn't be too long. You want a cup of coffee?"

"Sure, why not."

You're amazed at how fancy clinics are these days. Everybody's got an automatic coffee machine with a dozen different flavors. The Crenshaw Center is particularly fancy with huge fish tanks, reclining chairs, cozy little sitting areas, and muted green carpet with a leaf pattern. You notice the leaf motif is carried throughout the facility — probably a symbol for life. *Probably can't hurt.*

You see Wayne walking toward you with the coffee. It tickles you to see him bouncing off the tips of his toes. You laugh when you think of how hard you tired to get him to plant his feet on the ground when he was a kid. He would always plant his heels hard and then quickly roll off onto his toes with a little bounce. Oh, well. If that's the only thing you couldn't break him of, then you figure you did good enough.

"I got you a hazelnut, Mom. I got a hot chocolate."

"Thank you, son."

The two of you sip your drinks and after a long silence you speak. "Wayne?"

"Yes?"

"I love you."

"What? I love you too, Mom. Please don't worry."

Just as you are about to tell him you appreciate all he's been doing for you and that you want *him* not to worry, someone calls your name.

"Good morning, Mrs. Brewster. I am Parvesh Singh." A slender young man, in a short white coat with dark hair and eyes greets you.

You shake his hand. "Please, call me Gloria."

"And I'm Wayne. I'm the son."

"Ah, yes!" he replies, shaking Wayne's hand. "A pleasure to meet you both. Would you follow me, please?"

As you walk toward the elevator you ask, "And what shall we call you? Dr. Singh?"

"Oh goodness, no, not yet. Soon, it is to be hoped. I am a dental student. As you are Gloria, I am simply Parvesh."

"Well, pleased to meet you, Parvesh," you say, giving him a good Texas slap on the back.

You and Parvesh gab all the way to the examination room. Wayne, on the other hand, remains silent. He's obviously nervous. You find this a little weird. You seem to be handling the possibility of your illness much better than your son.

Once in the examination room, Parvesh tells you, "If you would change into one of those gowns, please? I will go see if Dr. Nguyen has arrived. We shall try to get this over with as quickly as possible."

"Quickly works for me just fine. My stories'll be on soon."

Parvesh leaves. You sit in the examination chair. The back of your thighs sticking to cold, black vinyl. "I hate these dang gowns. They never fit. They're ugly as sin and your barn door is always flappin' in the wind."

"Mom, this is a hospital, not Neimen Marcus."

"Well, then *you* wear it."

There is a knock on the door. Dr. Nguyen enters the examination room followed by Parvesh.

"Hello, again Gloria. Good to see you."

You like Dr. Nguyen. She has been quite thorough in your opinion and seems truly interested in you as a person, not just a patient.

"Howdy Dr. Nguyen. You remember my son, Wayne."

"Yes, nice to see you again, Mr. Brew —"

Wayne takes Nguyen's elbow and gently guides her to the door. He turns to you just before she and Wayne disappear into the hall. A somewhat surprised Parvesh follows them out. You hear Wayne from around the corner tell him, "Please close the door."

You hear muffled voices in the hall but can't make out what's being said. Minutes pass. *What is going on out there?* You look around for a magazine, or anything, to pass the time. There's a drawer under the sink and you it pull it open. *Ooh, lookey here!* You take out a gray stethoscope and try it on. You place it over your heart. *Yup, sounds just like in the movies.*

Time passes. They're still out there chatting. You can't take it any longer. "CAN Y'ALL KEEP IT DOWN OUT THERE? MAKES IT HARD FOR A WOMAN MY AGE TO SLEEP NEKKED!"

Seconds later, a nervous Wayne, grinning Nguyen and surprised Parvesh return to the room. Dr. Nguyen pats you on the shoulder. "Okay. So, today we will do a biopsy. It's no big deal. It takes five seconds."

You don't like hearing that at all. "I don't mind telling you, doctor, I had a conical biopsy for a cervical cancer scare a few years back, and I did not enjoy it one bit."

"I don't blame you," Nguyen responds. "I had one myself. No fun at all." She takes a tray of surgical instruments from a nearby cabinet, and sets it on a nearby table. You decide you don't want to look at them. Ngyuen begins pulling on latex gloves. "Unfortunately for us both, I left my magic wand at a dental convention last month. Will have to use this instead." Nguyen gestures with a long swab, "I'm going to brush some anesthetic over your tongue."

You dutifully stick out your tongue. Nguyen swabs the area around the lesion. When she is finished, you smack your lips. "That tastes awful."

"Odd. It's supposed to taste like piña colada."

"Well, I hate piña coladas. That might explain it."

"How's your tongue feeling?"

"It's getting numb. I guess I'm going to be talking funny for a while."

"Mrs. Brewster, I'm Vietnamese. To me, you all talk funny."

You laugh. You like her sense of humor. She makes you feel comfortable. You hope your grandson will make his patients feel this way when he's the doctor.

After a few minutes, Dr. Nguyen tilts your chair back into a reclining position. "I don't need that sucky thing in my mouth?" you ask.

Before she puts on her surgical mask, Nguyen smiles at you. "No. I don't want to take the chance of losing the sample down the tube. As I said, this will be quick."

Parvesh pulls on a mask and stands by her side. He picks something up from the tray, which you take as your cue to close your eyes and open your mouth.

"Okay, Mrs. Brewster …" Dr. Nguyen begins as you interrupt her.

"Caw me Gworia."

"Okay Gloria, I need you to remain still. You might feel a little pinch, because I'm not telling you anything new when I say the tongue is a big mass of nerves and — oh, never mind. I'm finished."

You open your eyes in surprise. "Reyee? I'm imprethed."

"As I said. You all talk funny." Nguyen works the foot pedals to return you to a seated position. "Here is 'the sucky thing,' if you require it. Let me write up my notes so Parvesh can enter them into the Crenshaw Center's fancy new EHR system, then

I'm going to walk this sample up to the lab. If you come with me, I'll show you where you need to go for your blood work. The CAT scan is close by."

"Hey, you got another one of these designer gowns?" The feeling is quickly coming back to your tongue. "I need to close up the backside."

"Yes, of course. I'm sorry. They're awful, I know."

Dr. Nguyen hands you another gown. You put it on like a robe over your other gown and feel a little more secure. As you tie the front strings, you wonder about the conversation that took place in the hall without you. *What was that all about?* You wonder if you should ask the "Gang of Three" about it. Then again, maybe not. Maybe you don't want to know.

TURN TO PAGE 319 TO ASK, "WHAT'S GOING ON?"
TURN TO PAGE 318 TO REMAIN SILENT.

# Gloria: Don't Ask

AS NGUYEN TURNS TOWARD THE DOOR, you pull the stethoscope from inside your gown. "These look like fun! Can I have 'em?" You place the stethoscope in Nguyen's hands, who returns it to a drawer under the sink.

"We would like to keep these for the next patient, Gloria. How about a lollipop instead?"

You break into a wide smile. "Deal."

TURN TO PAGE 320 TO CONTINUE.

# Gloria: Ask

AS NGUYEN TURNS TOWARD THE DOOR, you pull the stethoscope from inside your gown. "These look like fun! Are they only for listening through doors?"

You place the stethoscope in Nguyen's hands, who puts it back in a drawer under the sink. No one says a word, so you break the silence. "Y'all planning my demise?"

Wayne speaks. "Mom … um …" Wayne clears his throat. "We were just discussing the schedule of your tests for the morning, nothing more."

You're not buying it. "Biopsies are to look for cancer, right? Is that what you think I have?"

"No Mom, no one thinks you have cancer. This is just to make sure."

You funnel out the door with the group. Just before you part ways, Dr. Nguyen turns to you and looks you in the eye. "Yes, Gloria. Biopsies are a test for cancer."

TURN TO PAGE 320 TO CONTINUE.

# Gloria: Test Results

A WEEK LATER, you and Wayne return to the Crenshaw Center. Parvesh greets you when you arrive.

"Hello, Mrs. Brewster. Mr. Brewster. How is my favorite patient today?" He really is charming. So polite! *Too bad Stephanie isn't a couple years older.*

"Remember Parvesh, call me Gloria, and I'm fine, and I am eager to get this over with."

"I totally understand … Gloria. Please have a seat and I'll be right back with the doctor."

It's been a hard week, mostly for Wayne. You wish he wasn't such a worrywart, but he's been that way ever since he was a little fellow. You are convinced he inherited the worry gene from his father, rest his soul.

After a brief wait in the exam room, Dr. Nguyen enters. She is accompanied by a trim man in a business suit. Wayne looks at the two and points to the exam room door. "Can we …?" The two nod and step back out into the hall. Wayne closes the door.

When you hear the latch, you reach under the sink. The stethoscope is gone. *Dang!* Muffled voices outside the door slowly disappear. *They're walking away! I've got a good mind to open the door and tell them to be straight with me.* Except, part of you doesn't want to know everything. Not just yet. What's that saying? *Oh yeah.* "What you don't know can't hurt you." *Right?*

Wayne and Parvesh return. Wayne's face when he comes back into the room is a billboard. He has "that look." Whenever Wayne is upset or disappointed, he pushes the corners of his mouth down and closes his eyes half way. He also sinks his chin into his chest. Yep, the "look" is all over him. *Something is wrong.*

"Good news," Wayne says. "The docs say there's nothing to be worried about. They've just got to figure out the treatment and then you'll be cheating all your girlfriends at Uno again."

You take a deep breath, then let it out. "Oh, thank God. I *do not cheat* at Uno, Wayne. I *win*."

Parvesh hands you some paperwork. "We have to fill out some forms for you to go over and sign, Mrs. Brewster."

"Gloria."

"Oh yes, Gloria. Excuse me again."

You look through your purse for your glasses. "Dangit. I left my reading glasses at home. I guess you gotta read 'em to me."

"Certainly," Parvesh says and sits next to you. "This first form is call informed —"

Wayne pulls the papers from Parvesh's hand. "Tell you what, Parvesh, you're a busy guy. I'll take care of this. Can I just drop 'em off with the girl out front?"

"I suppose so. If there are questions —"

"Not to worry. We'll flag them and ask before signing. Is that okay?" Wayne's question is delivered as a statement.

Parvesh leaves the room. Wayne takes the stool out from under the sink and puts the forms on the table. He starts filling them out. You begin to feel a little helpless. It's more than just forgetting your glasses. For some reason, you're tired too. Worst of all, you are getting the distinct, and sinking, feeling that maybe things are a bit more complicated than anyone cares to admit.

TURN TO PAGE 323 TO CONTINUE.

# Gloria: Home

YOU STARE OUT THE WINDOW ON THE DRIVE HOME. You are amazed at the massive complex of buildings and walkways that make up a large medical center. It seems like a small country with its own language and culture. Patients are just visitors from foreign lands.

You think about striking up a conversation with Wayne, but don't think your words would penetrate the heavy silence. It breaks your heart to see him so unhappy. You would do anything to help him if you could. *What mother wouldn't?* "Don't worry, son. Everything will work out. It just takes time. Maybe you and Sheila will patch things up. If you can't, you can't. Either way you'll be fine."

Wayne takes a deep breath and lets out a heavy sigh. He responds with an affirmative head shake.

As the car pulls into the driveway you see your grandson, Walter, sitting on the front porch. It's a pleasant surprise. He rushes to the car and opens your door.

"Hey, Nani!"

"Hey, Buddy. I am *sooo* glad to see you." You pull him to you and give him a hug and a little kiss on the cheek. You've always had a special relationship with Walter. You were the one he always confided in first — or asked for advice. Not that Wayne isn't a good father. He is. It's just ... it's just that most kids need someone other than their parents to be their confidant. You are glad that Walter chose you.

"So, what did the doctor say?" Walter asks.

Wayne jumps in, stuffing papers under his arm. "They said it's nothing."

"Can I see those?" Walter slides the test results from his father's armpit.

"Hey! Gimme —" Wayne lurches forward.

Walter steps away and puts his hand up as a stop sign. As he reads, he frowns. "What do you mean, *nothing?* It says Nani has a 3 cm lesion that —"

"That's enough Walter ... I ... I ..." You know that stutter. Wayne always stutters when he's hiding something.

"Wayne? Are you hiding something from me?"

"What did they tell you, Nani?" Walter asks.

"Walter! I told you to *stop* interrogating her!" Wayne's stuttering is gone.

"*Boys!* No fighting." Time to break this up. You look at Wayne, then at Walter. "I want to know what those things say. Go ahead, Walter, you're the doctor. You tell me."

"Mama!" Wayne tries to stop the conversation.

"Not another word out of you, Wayne Brewster. *Do you hear me?*" You are surprised at your own raised voice.

Wayne falls silent. Walter continues.

"It says ..." Walter stops and looks deep into your eyes. "It says, Nani, you have the beginning of cancer."

You feel like you just got the wind knocked out of you. Walter takes your arm.

"Nani, they caught it at the beginning, so that's good." Walter pulls you close. "It's gonna be okay. Come on, let's sit on the porch. I'll get us a cola and we'll look over these papers." He escorts you to the old pink porch swing where you and he have shared many a Dr. Pepper and good conversation while rocking. You remember a time when his feet just barely hung over the edge. He'd want you to swing fast, but you would always tell him, "This here is for talking and thinking. You want to swing? Go on down to the park."

As you walk, arm-in-arm, you whisper. "Walter. There's a box of chocolate turtles on the fridge. Bring them out too."

Walter sits you down, then returns with drinks and candy. He places the box between the two of you. His feet rest firmly on the ground now. Wayne sits on the bottom stoop of the concrete steps. He picks at weeds growing through the cracks.

You open the box and offer Walter a turtle. "Wayne, you want a turtle?" He shakes his head, no.

Walter reads through the papers slowly, giving you a synopsis along the way. "It says that you are enrolled in a study?"

"I am?"

"Yes, you signed up today."

"I did?"

"Yes. Didn't they go over the details of the study with you at the clinic?"

"No. Who was supposed to?"

Wayne plucks a weed.

"Dr. Crenshaw, or maybe one of his assistants."

You look at your grandson in confusion.

"Nani, there is something called 'informed consent.' It means you have be fully informed, and you have to make your own decision about signing up for a study. It looks like neither of those things happened today."

"But, it's supposed to be a miracle drug." Wayne says.

"Dad, it doesn't matter. Informed consent is a requirement."

"What do I need a miracle drug for?" Now you are getting upset.

Walter gives Wayne a dirty look. Wayne goes back to picking at weeds. Walter continues with his summary.

"It says you could experience some side effects. Hmm … I don't like that."

"They said it was the best option, damn it!" Wayne grits his teeth.

"I can't believe you would sign Nani up for a study, Dad, without telling her what she was getting in to. It's not right!"

"We'll what am I supposed to do, Walter. I didn't want her to worry. Thanks to you, Nani *is* worried. You blew it!"

"*I* blew it? *You* blew it! You *always* blow it!"

"That's it! I'm outta here."

"Good!"

You hate to see them argue, but you're relieved that Wayne is leaving. The tension between father and son was getting unbearable. With Wayne gone, you might be able to get to the bottom of what is going on with you.

"I'm sorry, Walter. I didn't mean to cause such a ruckus."

"No, Nani, you didn't cause anything." Walter pats you on the shoulder.

"Okay. It says there are some serious side effects. You could get weak and experience nausea. You could get anemic. You hair could fall out." He flips a page. "Well, there are many things in here. It says that you will be in the study for at least six months. You will be taking the drug in pill form and you will have to come in every month for blood work and other tests." Another page flip. "There are no guarantees that this will work, although the drug seems to be promising based on previous studies."

"Poppy suffered so much when he got sick. I don't want to go like that, Walter."

"I know, Nani. That was a long time ago. We know a whole lot more about cancer now."

"Mirabelle Wiggins suffered like hell too."

"I didn't know Mrs. Wiggins has cancer."

"*Had* cancer. Breast cancer. They got it too late. Spread everywhere. She died a few months ago."

"Well, it looks like they got yours in time. It's a Stage I."

"I remember them saying Mirabelle's was Stage III. I guess that's what got her."

"There's a big difference between Stage I and Stage III."

"Well, you're the one who's going to be the doctor."

You and Walter sit on the porch swinging back and forth until night fall. The Austin night air is crisp and clean and quiet. Walter brings your lavender shawl from the living room. It's a perfect night, sort of, except for the situation. Walter looks up the miracle drug right there in front of you on his iPad thingamajig. This is one time you are grateful for technology. He tells you that, apparently, the initial reports about the drug are very encouraging. Dr. Crenshaw himself is running the study. He knows Dr. Crenshaw from the school and tells you that he has a good reputation.

The two of you discuss other treatment options. It seems that you could choose to go the traditional route, which means surgery followed by chemotherapy, or you could just take the experimental drug and not have to go through surgery. Either way there would be side effects.

"You know, I don't mind dying, Buddy. Don't get me wrong, I want to live. I am hoping to see you graduate and maybe even see you married with kids of your own. I know I don't have any control over any of this. Only the man upstairs knows what's in store. Looks like I am dammed if I do and dammed if I don't."

"I know this isn't easy, Nani. You know that I will be there for you. No matter what you decide, I will always be there for you."

You notice Walter wiping a tear from the corner of his eye. In the still of the night, after a long silence, he places an arm around your shoulder and whispers, "I love you, Nani."

TURN TO PAGE 333 TO JOIN THE STUDY.

TURN TO PAGE 329 TO DECLINE PARTICIPATION.

# Gloria: Conventional Treatment

NOW THAT YOU KNOW YOU HAVE CANCER you want to get all the facts. You have a meeting scheduled with Dr. Nguyen to go over details of the study and other treatment options. Walter has decided to join you and Wayne for the meeting. Having your grandson with you makes you feel secure. You know he will be honest with you when others are not.

"Hello, Gloria." Dr. Nguyen seems happy to see you. "Come on in and have a seat. Hello again Mr. Brewster. And hello …?"

"Dr. Nguyen, this here is my grandson, Walter — and he's one of you!"

"I beg your pardon?"

"I'm a med student, third year."

"Good to meet you."

"It's good to meet you too, Dr. Nguyen."

You take the floor. "I've got some questions for you, Dr. Nguyen. I want facts, not sugar coating. Tell me what I need to know so I can make a decision."

"I agree completely. You need to know all your options and —"

Wayne jumps in. "The Kuritin study is her best option, right, doctor?"

"Not necessarily, Mr. Brewster."

Wayne seems surprised, and maybe a little irritated with Dr. Nguyen's response.

Nguyen continues. "Normally we would treat this kind of cancer with surgery. Sometimes we use radiation if the lesion is not easily accessible. You're in Stage I, which means it has not yet spread. Radiation has side effects. Surgery has all the normal possible complications that come with surgery, plus a risk of permanent disfigurement. There is just no way to get around that. Of course there could be reconstruction surgery down the road."

"What about Dr. Crenshaw's study?" Walter asks.

"I've heard it has possible benefits, but the results have not yet been published because the study is ongoing. We don't know yet how it will turn out. If it turns out to do what it is supposed to do, you may not need surgery or chemo."

Walter again. "*May* not?"

"Yes, *may*. Again, we don't know the results yet."

"That's not what I hear, doctor." Wayne is pushing.

"Of course, if Kuritin works as advertised — and I say, *if* — it would be a major breakthrough. Right now, the KURE study has a couple of years left to run, and then we have to analyze the results. That could take another couple of years. Then, others will try to duplicate the results, and —"

"My mother may not have that much time, Dr. Nguyen."

"Mr. Brewster, I'm just telling you that you — that is, Gloria — should keep in mind there are years of data on standard treatments, but little on Kuritin."

"Don't you think she should go for the study?" Wayne is trying to get Dr. Nguyen to agree, but Nguyen won't be manipulated.

"It's Nani's decision, Dad."

"Thank you Walter. Give me a minute to think, boys." You've never been much of a gambler, unless you're playing Uno. Seems like the traditional approach is the bird in the hand. You don't like the idea of chemo, but you like the idea of being a guinea pig even less. You decide to move forward with traditional treatment.

"I think I'm gonna pass on the study, for now."

"Oh, God. Mama!" Wayne is almost shouting.

"Quiet, Wayne." This is *my* decision.

* * * *

It's been four years since that fateful day in Dr. Nguyen's office. You remember oh so clearly how the doctors described how oh so treatable Stage I cancer was. "Not to worry" was the tone of most conversations. The doctors were still hopeful and optimistic when they announced your surgery for the oral lesion was successful, but that they had found two other small lesions in your lung. Unfortunately, they said, these were Stage III.

Chemotherapy over the past few years has left you a shell of your former self. You can't hold a paintbrush. You can't play Uno. You rarely leave your bedroom.

Lately, you hear precious little hope in the doctors' voices, especially since last month when your were described as "Stage IV."

Last week, you overheard a nurse say something about "end stage." You plan to ask Walter what that means the next time he visits.

TURN TO PAGE 323 TO TRY AGAIN.

# Gloria: Join the Study

NOW THAT YOU KNOW YOU HAVE CANCER you want to get all the facts. You have scheduled a meeting with Dr. Crenshaw to go over the details of the study. Many things are going through your head. *Odd that they call the study, "KURE." I'll just bet that's an abbreviation for some gobbledygook, but it sure sounds good.* Walter has decided to join you and Wayne for the meeting. Having your grandson with you makes you feel secure. You know he will be honest with you when others are not.

Upon your arrival you are greeted by the guy you've come to understand to be the head honcho 'round these parts, Dr. Crenshaw. Parvesh flanks him to the left and one step behind — *his faithful soldier*, you think.

"Well, hello Mrs. Brewster. I am Dr. Crenshaw. I've heard a lot about you." You decide to let him call you "Mrs. Brewster." Something about him makes you want to keep things formal.

"Happy to meet you too, Dr. Crenshaw. I recognize you from your portrait." *Hell, it's hard to miss*, you think. *It must be eight or ten feet tall and must have cost a fortune to frame that thing.* You figure that's why they made it the first thing you see when you come in the door. *Too bad it's not well painted.*

"I am afraid that it's not me. It's my grandfather, although people say we look alike. Please, come in." He escorts you and your family into his office.

"This here is my son, Wayne."

"Yes, we've met. Good to see you again, Mr. Brewster."

"This is my grandson, Walter. He's one of you!"

"Oh?"

"I'm a medical student, sir." Walter shakes Crenshaw's outstretched hand.

"Your grandmother is quite right to be proud. So, how's medical school treating you?" Crenshaw's voice seems feigning interest.

"Very well. Thank you, sir."

The "office" is unlike any room you have ever seen, except on the old *Dallas* television show. It looks like it belongs to the CEO of a big oil company. Next to Crenshaw's large wood paneled desk are a sofa, arm chairs, two side tables, and a coffee table. They must be antiques, you assume. On the other end of the room is a beautiful, and very long, modern credenza, adorned with a dozen gold statues and glass awards. Above it are photos of Dr. Crenshaw shaking hands with Presidents Obama, Bush (junior and senior) and Clinton. There are others with Grandpa Crenshaw shaking hands with Reagan, Carter, Nixon and — I can't believe it — John Kennedy. *Both Crenshaws were dashing men in their youth*, you can't help but notice. In the middle of the room is a large round conference table, again solid wood. A silver tea set is sitting in the middle of the table. Chairs are everywhere. Certificates and diplomas adorn every wall. Crenshaw's window to the outside frames a private garden, immaculately kept. To the right of the window is a gorgeous, probably old, probably expensive, grandfather clock, ticking loudly. *Somehow, all this just doesn't seem right*, but you're not sure why. *That's peculiar.* His desk looks like no one works here. *It's spotless!*

"Would you like something to drink?" Crenshaw asks.

"Yes, I'll have a vodka gimlet." Everyone laughs.

"The closest I can come to that is lemonade. Will that suffice?"

You accept the offer. Parvesh retrieves drinks from a small refrigerator hidden inside the credenza.

"Now, I know you have some concerns, as one should when considering any pharmaceutical trial for a new drug. I want to assure you that this study has shown wonderful results. We have been *very* successful. For many people, it's been a miracle."

"That's what I've been telling her," says Wayne.

Walter offers another perspective. "Dr. Crenshaw, um, I have been reading up on the study, you know, to help my grandmother understand it a little better? I think her main concern is the side effects and what her options are if the drug doesn't work … that is, if for some reason —"

"Side effects are minimal, something like having a mild case of the flu. All are managed with medication for symptom relief. No big deal. However, if for some unforeseen reason your grandmother does not do well with the medication we can always treat her with traditional protocols." He pauses in dramatic fashion. "Chemotherapy and surgery."

The thought of chemotherapy upsets you. "Lord, I don't want chemo. I've seen too many family members and friends go through that. In the end it didn't do a thing."

"Sometimes it works, Nani. Often times."

"He's right, Mrs. Brewster." Crenshaw smiles at Walter. "It looks like you're paying attention in class, son. I'm not worried about you having chemo, Mrs. Brewster. I don't think you'll need it. I believe … I *know* … Kuritin is the right thing for you."

You want to believe Dr. Crenshaw badly. He certainly seems confident. If worse came to worse, you figure you could always go back to the traditional route. That's a lot more scary, but you know Walter and Wayne will be there to see you through it. Still … it seems like the right thing to do now is to go ahead with the study.

"It's your decision, Nani." Walter takes your hand. "Only you can decide what's right for you."

The room is totally silent, except for Grandpa's clock. "Tick … tick … tick …" All eyes are on you. The clock gongs.

"Where do I sign?"

Wayne looks like he was just about to pass out.

TURN TO PAGE 337 TO CONTINUE.

# Gloria: First Month

THE DAY HAS COME FOR YOUR FIRST TREATMENT. You are relieved and anxious at the same time; kind of glad the treatment is starting, sort of worried it might not work.

You and Wayne show up in the Crenshaw Center a few minutes early. While you wait, you enjoy a cup of hazelnut coffee from the fancy coffee machine.

"Are you sure you should be drinking that before your blood work, Mama?"

"They didn't say I couldn't, Wayne. I know you're worrying about me, but I'm okay."

"It's a lot to deal with, Mama."

"You've got too much on your plate, Wayne. I appreciate you helping me, but I want you to pay attention to you. You're going through a divorce. You're on probation at work. You don't need to be worrying about me. The best thing you can do to help me is to take care of yourself ... okay?"

"I'll try, Mama."

"Okay then."

You finish your coffee and close your eyes. With a deep breath you sink into a jade-green waiting room chair. You feel like taking a nap. It would be easy to do in this clinic. Your chair's back has a "give" to it. You can request a blanket anytime you like. It's almost like being in a spa, except, of course, it's not a spa.

You forgo the blanket and nod off. You feel at peace. Weightless. As if you are floating in the clouds. You're at your easel, painting. Broad brush strokes cover the canvass. Someone calls to you in the distance "Gloria …? Mrs. Brewster …?" In his voice is a sweet familiarity. "Gloria … Hello?" You feel a gentle pat on your hand. "Hello?"

You slowly open your eyes. You are groggy. Your eyes are barely open and you can't help slurring your words. "Hello Parvesh."

"It's time. We're ready for you."

You stand up with Parvesh and Wayne's help and hold on to their arms all the way to the examination room. You feel tired and old. It feels like you've aged ten years in the past two weeks. You saw this happen to Poppy and Mirabelle. They got diagnosed with cancer, then bang, they started using a walker. Not long after that, they're using oxygen tanks. Before they knew it, they're dead. *Bang, bang, bang.*

*Oh Hell,* you think. *I'm not dead! I'm just sleepy. Get a grip old girl!*

"Okay boys." You stiffen up. "Let's get on with it!" *That's better. Much better.*

Parvesh takes your blood pressure and sticks you for the blood draw.

"Make sure you leave me some, Parvesh. I need energy to play Uno this afternoon."

"Maybe they want you to rest, Mom."

"It's okay Mr. Brewster. She can resume normal activities. That's one of the benefits of this therapy. It allows you to live a relatively normal life."

You hear a knock on the door. It's Dr. Crenshaw. You are starting to think that he looks like a movie star — *an older version of someone ... * Dr. Kildare. *Yeah, that's it.* He reminds you of an older Dr. Kildare. Crenshaw comes in tapping a card of blister-packed capsules. "This is the miracle we're working on," he proudly announces. "All you have to do is take one of these every day after breakfast. Then we'll see you again in a month and see what kind of progress we're making."

You take the pack from his hands, "Your lips to God's ears."

You look at the milky white capsules. "Not real purdy, are they? They look like the worm medicine I used to give my dog."

Crenshaw shrugs. "What can I say? Finding the most attractive colors for the capsules is another research group entirely."

"Okay, doctor, I'll give you a pass this time. Should I take one now, or —"

"Tomorrow. Right after breakfast."

"I can do that, doctor." You turn to Parvesh. "You need to poke me anymore?"

"Not today, ma'am. I think we've taken quite enough from you."

"Okay then." Wayne helps you down from the table. "I will see you gentlemen in a month." As you head toward the door, you lean into Parvesh and whisper, "And, it's Gloria, not 'ma'am.'"

"Yes, ma'am — Gloria."

Leaving the clinic, you turn to Wayne. "How ya feeling son."

"Better."

"Me too. How 'bout we go get a soft serve at McDonald's.

"Sounds good, Mom."

✳ ✳ ✳ ✳

Five days later, you call Wayne early in the morning. "I'm real sick, son. I need your help."

"Oh, God. I'll be right there, Mom."

You feel so weak, you can't even pick your head off the pillow. You have never felt this helpless in your life.

Wayne makes it to your house in record time. You want to scold him for speeding, but you don't have the energy.

"What happened, Mom? Did you fall? *What*?" He is frantic.

"I don't know. I've been up all night, vomiting."

"You've been drinking anything?"

"A little 7-Up. Can't keep it down. Tried Saltines. No good either."

"I'm gonna call the doctor, right now."

You are relieved. Being at home alone like this has been frightening. You close your eyes and pull your pink chenille bedspread to your cheek. You love the old bit of cloth. It belonged to your mother. The tufted yellow and blue flower pattern reminds you of happier times. It's funny how a simple thing can offer such comfort.

"My mom's been up all night vomiting." You can hear Wayne from the kitchen. "She is not keeping anything down.

She tried 7-Up and Saltines. That's what she always gave me when I was sick. I don't know what else to do. I'm wondering if she should even take her pill this morning, Parvesh."

While Wayne is on the phone you hear someone calling from the living room. It's Walter. He enters the bedroom.

"What's going on, Nani? Stephanie called me all worried. She said dad was rushing over here."

Wayne comes back from the kitchen. "She's been vomiting all night. Can't keep anything down. I called Parvesh. He says he is going to call Dr. Crenshaw to see what he wants to do."

Walter places the palm of his hand on your forehead and puts his hand on your wrist. "You definitely have a temperature. Your heart rate is a little fast, which is normal under stress." He gently pinches the skin on your forearm. "You're getting dehydrated, so we better try getting some fluids in you. Dad, bring Nani a glass of water." Walter brings the glass. "All you have to do is take little sips, Nani. Let's wait for the doctor's call before we try giving you anything for the fever."

Just the thought of drinking water makes you want to vomit. You know it's important, but you would welcome an IV right now.

The phone rings.

"Dad, put it on speaker, will ya?"

"Hello?" Wayne answers.

"Hello, Mr. Brewster. It's Dr. Crenshaw. I understand your mother is not feeling well."

"Yeah, that's right. She's been vomiting all night and she can't keep anything down."

"She has a temperature. Hi, Dr. Crenshaw. It's Walter Brewster."

"Making house calls already are we? I'm sorry Mrs. Brewster isn't feeling well, but there is no need to worry. The drug she is taking is very powerful. Initial reactions like this are common. Her body is undergoing a period of adjustment and you shouldn't be worried about it. Keep her on the regimen. You can give her some chicken soup or Gatorade before taking her pill. After she keeps that down give her some Tylenol for the fever. You could try the liquid kind if the pills are too difficult to swallow. She should be feeling better by the end of the day. If not, call me back. We may need to see her."

"Is there anything else we should be concerned about, sir?" Walter asks.

"No, son. Everything will be fine."

You want to say, "Thanks," but all that comes out is a cough.

"Grandma says thank you."

"You're welcome. Feel better now, Mrs. Brewster. Goodbye."

Walter sits by the bed holding and rubbing your hand. He takes the cup of water and holds it for you while you take sips, which requires all the energy you've got. Your head sinks into your pillow. The sun shines through the cracks of the Venetian blinds. You can see little specs of dust dancing in the beams of light. The movement is soothing. You watch until you fall asleep … the first time you have slept all night … a deep sleep … a kind sleep … a …

TURN TO PAGE 344 TO CONTINUE.

# Gloria: Soul Searching

IT'S BEEN SEVERAL DAYS since your reaction to the drug. Dr. Crenshaw was right. You are feeling a little better, but not yet well enough to want to play Uno with the girls. Or paint. Walter has been staying with you and that's been wonderful. Your family has been very supportive. Your friends visit and bring food. You appreciate everything, but you just don't like living this way.

You begin to think that perhaps you should quit the study. For the first time in weeks you and Walter sit outside on the porch swing.

"Buddy, I don't know if this is the right thing to do."

"What do you mean, Nani?"

"The study. Maybe we should quit."

"Maybe."

"What do you think, Buddy?"

"Are you feeling better?"

"Better than I was, but not good."

"Well, if you do the chemo and surgery you're still gonna be sick, but maybe it would be a better kind of sick. I just don't know. With surgery? You may wind up a little disfigured. I don't know how you'll feel about that. You probably won't be able to eat much while you're healing from the surgery.

"I'm not eating much now."

"I don't know, Nani. I don't know what to tell you."

At this point you want it all to stop. You are sick and tired of feeling sick and tired. You have never felt so low in your entire life. You feel hopeless and beaten down. You are surprised that your will to live is hardly there. You begin to think it might be better to just give up altogether. Your eyes give your thoughts away.

"I know this is hard, Nani. But, I think you've got a great chance to beat this thing. You're in the first stage of the disease. There is an extremely high success rate with treatment in this stage. I think it would be a shame to give up now."

"Yeah, maybe so, but I am beginning to think that leaving it to the man upstairs might be just as good, maybe even better.

Walter looks at you quizzically.

"We'll see, Buddy. Don't worry. I'm too mean to give up. You know that. I'm tired now. Let's go back in the house."

Walter helps you get back in bed. Sitting outside sapped all of your energy. You know that in some way, this is just as hard on Walter as it is you. He's young and full of life. He's studying to be a doctor. He's training to do everything he can to save lives. He doesn't have the experience to know that, sometimes, some lives can't be saved.

You close your eyes and keep thinking through your options. Walter sits dutifully by your side in the old, white, wicker rocker that your mama used to rock you to sleep and that you used when Wayne was a baby, and then again with all your grandkids. You think about how one day it will belong to Walter, who will probably do the same thing. *Hell, I want to get better, but the treatment is taking a toll on me. I can't do the things I love to do, so why bother? Yet, maybe the worst is over. Maybe I just gotta hang on a little while longer until I turn the corner. Maybe I ...*

Your eyes are getting heavy. Walter has fallen asleep in the rocker. The warmth of the afternoon sun fills your bedroom. You pull your bedspread to your cheek and take a well deserved nap. Maybe things will be better when you wake.

TURN TO PAGE 347 TO WITHDRAW FROM THE STUDY.

TURN TO PAGE 349 TO STAY ENROLLED IN THE STUDY.

# Gloria: Leave

THE NEXT MORNING you feel worse. Walter brings you some oatmeal and a little warm milk.

"How ya feeling, Nani?"

You just shake your head. You try to eat something but all you can do is move the oatmeal around in the bowl. "I want you to call Parvesh and Crenshaw and tell them I'm not gonna continue with the study."

"Are you sure that's what you wanna do?"

"Yeah. I'm sure."

As soon as the words come out of your mouth you feel a little lighter. You pick up the oatmeal and take a bite. It tastes good and comforting. You take a small sip of warm milk. That tastes good too. Walter is in the living room talking to Parvesh.

"Yes, I'm sure. That's what she wants to do ... No, we don't need to talk to Dr. Crenshaw. Thank you ... I'll tell her you wish her well. Thank you, Parvesh. See you around campus. Yes. Goodbye."

Walter returns to your side. "You are free."

"You don't know how true that is."

A few hours later, Walter leaves for class. Moments later, the phone rings. It's Wayne. "Mama, you can't quit!"

"Wayne, I'm not quitting, I'm choosing a different path. I am just —"

"No, you're quitting the trial. I gotta call from Dr. Crenshaw. I know that you are quitting the trial and I'm not going to let you."

"Wayne, I —"

"Mom, Dr. Crenshaw is concerned about you. You have to stay in. You *have* to. For us. For me, for your grandchildren, for all of us. You have to stay in the trial. It's your best hope. Please. *Please*."

You are starting to cry. "Wayne ... I ..."

"Please Mama. *Please?*"

"Okay. Okay. Call Crenshaw back. Tell him I've changed my mind. I'm not quitting."

Turn to page 350 to continue.

# Gloria: Stay

THE NEXT MORNING you feel better. Walter brings you some oatmeal and a little warm milk.

"How ya feeling, Nani?"

"A little better. Guess I needed the rest." Things don't seem quite so bleak this morning. You pick up the oatmeal and take a bite. It tastes good. You take a small sip of warm milk. That tastes good too.

"What are you thinking, Nani? About the study and all." Walter is obviously worried.

"Well, Buddy. I think we are going to give it the old college try. What do *you* think about that?"

"I think that sounds real good. Real good, Nani."

TURN TO PAGE 350 TO CONTINUE.

# Gloria: Second Month

THE MORNING FOR ROUND TWO HAS ARRIVED. You wake up to the phone ringing. It's Walter. "Good morning, Nani."

"Hi, Buddy." Your voice is weak.

"You remember what today is?" he asks.

You figure you're doing well if you can remember your name. *Today?* "No, I don't."

"It's time to go back to the Crenshaw Center to see how you're doing. I'm gonna take you."

"Oh. Okay."

"I'll be over around noon. Will that work?"

"Yes." You hang up the phone and fall back to sleep.

Walter and Wayne know that you have not been doing well. While you are no longer vomiting, you feel horribly weak most of the time. This is not what you, or they, were expecting.

Walter arrives to find you still asleep. "Nani? Hello?"

"Oh. Hi, Buddy. Sorry. Must have dozed off. Give me a hand will ya."

Walter helps you to the bathroom. You look in the mirror and are shocked by who, or what, you see. *Is that me?* Your skin is pale and your cheeks sunken. You try to powder your nose, but can barely hold the compact. It takes all you've got to brush your teeth. It even hurts to comb your hair. You do your best to get dressed. With Walter's help you make it to the car and are off to the Center. You haven't been in a car for over a month. The movement is unsettling.

It's a long walk to the Crenshaw Center clinic. Fortunately there are wheelchairs strewn everywhere, like a used car lot. Walter valets the car. A young man in a bellhop-like uniform helps you into the wheelchair. *I can't complain about the service around here, that's for darn sure.*

On the way to the clinic you notice how you are starting to look like everyone else around here. You hunch in your chair like they do. You look pale and ghostly like they do. You hope that the doctor will have some good news, like they do. You hope all this suffering has been worth it, *for all of us.*

Walter signs you in. Today you will have no fancy coffee. You will just sleep in that jade-green chair until they call for you.

Parvesh greets you as always. You've come to depend on the comfort of his sweet demeanor. You can tell that he is disheartened by your appearance. Still, he tries to be cheerful.

"Hello, dear Gloria. I am so glad to see you. I have missed your smiling face."

You have just enough energy to smile and nod. Parvesh seems to get the message.

"Hi, Parvesh. She's a little weak."

"Yes. I see. I am sorry you don't feel well, Gloria. Come, let us see what we can do."

Parvesh leads you to the exam room. He takes your blood pressure and the usual vial of blood. You notice that he is particularly gentle with you, and you appreciate that.

Crenshaw enters. "How's my favorite patient?"

You muster enough energy to say, "Feels like crud."

"So, what's going on?"

"Look, my legs are swollen."

"Yes, edema is normal. It should go away."

"I am still nauseous and I gotta tell you, I am having a hell of a time peeing, pardon my French, Parvesh. That's not a problem I've had since giving birth, you know?"

Walter speaks up, "And that's worrying me. Wasn't there a record of kidney failure in the early trials?"

Crenshaw shakes his head. "In early *animal* trials. That's why animal experimentation is important, but we're dealing with a different version of the drug here."

"Doc," Gloria says, "I don't know how much more of this I can take. I've had friends go through chemo, and this seems just as bad, if not worse. I'm not sure I can stay in the trial."

"That's probably the most common thing we hear in the cancer field — that the cure is worse than the disease. I know this sounds backwards, but the symptoms you're experiencing are proof that the drug *is* working. You may not feel like it right now, but you are making incredible progress with Kuritin. So much so that I feel safe reducing your dosage, and that should increase your comfort level as we continue your treatment. How about it, Mrs. Brewster? Can you give me another month?"

You sigh. You want to believe him. "I'm making good progress?"

"Absolutely."

"Well, I guess I'll keep going. I hope I start feeling this 'progress' you're talking about soon. And by the way, call me Gloria, will ya?"

"Okay, Gloria. Great. Parvesh will finish up here while I arrange for the reduced dosage."

You smile and go back to nodding. *I sure as hell hope Crenshaw is right.*

Dr. Crenshaw returns with the new dose. It's the same white pills with no personality. The instructions are the same: one pill with breakfast every morning. Call if there are any problems.

Parvesh reaches down and gives you a gentle hug. You appreciate the gesture. You reach up and pat him on the head.

Walter and Parvesh exchange a firm, long handshake. Parvesh offers, "If I can do anything, anything at all, please call me."

"We will. Thank you, my friend."

As you and Walter make your way through the clinic you make eye contact with fellow patients. You exchange half smiles and little nods. No one says a word, but you know what they are thinking. *I understand. Me too. Good luck.* You notice that the caregivers are also exchanging glances and nods. You suppose they are thinking the same things. It's only one disease, but it affects everyone.

TURN TO PAGE 354 TO CONTINUE.

# Gloria: Adverse Event

THREE DAYS LATER you are vomiting all over the bathroom again. You call Wayne, but he is not home. Walter answers instead. "I'll be right there, Nani."

You try to make it back to the bed, but your legs give out from underneath. You drop the phone and pass out on the floor. You wake up to Walter carrying you back to bed.

Walter immediately picks up the phone on your dresser and starts talking to someone. You can't tell what he's saying. Then, it's déjà vu. Walter takes out his cell phone and calls Parvesh. Parvesh listens. Parvesh calls Crenshaw. Crenshaw calls here.

"*I don't like this at all, Dr. Crenshaw.*" Walter is almost shouting. "This is not like last time. It's much worse. I've already called —"

Your eyes roll back into your head and you start convulsing. You can hear Walter frantically yelling, "Nani! Nani!" You try to speak, but can't.

Your entire body is trembling. You feel as if you are sweating and freezing at the same time. You have a taste in your mouth — like aluminum. You can't catch your breath. In the blackness, you hear Walter calling to you. You try to say something but the words won't come out of your mouth.

Now you hear a siren wailing to a stop. People are talking over cell phones and walkie-talkies. Strangers are calling you. "Mrs. Brewster? *Mrs. Brewster?*"

Your whole body lifts off the ground. No matter how hard you try to hold on, you can't. Something keeps lifting you higher and higher until you hear a loud *crack*, like a huge bolt of lightning. Then a *clunk*. You start dropping, fast. You can't stop. Your arms and legs are flailing in an icy, cold wind. You're dropping faster and faster until … *bam!* You're laying flat on your back. Your breath jumps right out of you. Everything stops.

A soothing, high-pitched hum fills the empty space as a swathe of bright light envelopes you. In the middle of white mist emerges a familiar face. *Mother?* It's mother, exactly as you remember her when you were a child. She's smiling. She's happy. There are other familiar faces too. Familiar, except, you can't make out who they are. You think one is Poppy, your husband, but you can't be sure. They are all smiling. They are all happy.

Your sense of time and place is gone. You are floating, suspended in a peaceful place for what seems an eternity. Then, you slowly back out of the cloud. Your Mother is getting smaller and the others. *Not yet. It's not your time, yet.* They are speaking, but their mouths do not move. Fade to black.

Way off in the distance, you hear a tiny beep. After several seconds you hear another, and then another. The beeps become louder and more frequent. Your head begins to throb. You push open one eye. You look around and realize you are in a hospital. Your throat is dry as a desert. You smack your lips together.

"Hi, Mama. It's Wayne. It's okay, you don't have to speak. You've been asleep for a while. You want some water?"

You see Walter holding your hand. Walter is smiling. You try to speak, but something is caught in your throat. You nod instead.

Wayne gently places a sliver of ice on your lips. It feels soothing and satisfying. You close your eyes and look for that soothing white light, but it is nowhere to be found. You become aware that it won't be coming back, at least not for a while. You're not sure if that's a good thing, or a bad thing.

TURN TO PAGE 357 TO CONTINUE.

# Gloria: Aftermath

AFTER FOUR DAYS IN THE ICU followed by two days in a regular room you are starting to feel better. All the tubes and monitors are gone except for one IV, "Just in case." You have been told that all you are waiting for are the doctor's orders so you can go home.

"I'll be glad when you're out of here, Mama," Wayne sighs.

"You ain't just whistling Dixie, son."

There's a ruckus outside your room. You assume it's Dr. Crenshaw. Every time he shows up it's like the circus coming to town. Most of the time, he's trailed by an entourage of residents and fellows. The only thing missing are TV cameras and commentators.

Dr. Crenshaw cracks open your door just wide enough to pop his head in. "Good morning, Mrs. Brewster."

"Gloria," you remind him.

"Right. Gloria, are you ready to go home?" he smiles.

"Just say the word and I shall be free."

"Then, you are free."

"*Hallelujah!*"

"Er, doctor, a word?" Wayne pushes Crenshaw back into the hall, but you're having none of it.

"*Hey! You two!* No more talkin' behind my back! *Do you hear me?*"

"Yes, Mama."

"Yes, Mrs. Brew — Gloria."

Wayne pulls the door open and Crenshaw enters. Crenshaw closes the door on the entourage behind him. The shock on Wayne and Crenshaw's faces is worth a photograph. *Wished I had one those them dang smart phones with a camera.* You figure you should have spoken up a long time ago.

Wayne speaks first. "Doctor … all that I was going to ask, is … we still don't know what made my mother get sick like this, do we?"

"No, we're not sure, but one thing I can tell you is that it is unrelated to the study. So you don't have to worry about that."

"You mean she can continue with the Kuritin?"

"Absolutely. She can continue with her medicine today. Not a problem."

Your heart dips as Wayne answers back. "Good. I was worried about that."

"The discharge orders are being written up now." Crenshaw turns to look at you. "Gloria, you will be out of here in no time."

"Thank you doctor. I'm looking forward to real food."

Wayne still has that shell-shock look on his face. "Yes, Doc. Thank you for everything."

Crenshaw tips his finger to an imaginary hat just before leaving. "You are both welcome."

The moment the door shuts, Wayne's face turns sour.

"What's wrong, son?"

"Did you hear what he said? It's *not* related to the study."

"That's a good thing, isn't it?"

"Well, yes and no. This has been an expensive stay. Medicare only pays eighty percent. You don't have a supplemental policy, Mama. If it's not related to the study, we gotta pay the remaining twenty percent out of our own pockets. That's a hell of a lot of money, Mama. We're talking thousands of dollars."

"I have some savings, Wayne. You have enough to worry about. Let me worry about my bills." You hate to see Wayne upset over this. He doesn't need anything else in his life to worry about. Hopefully, when you get home things will seem much better.

"I can't let you do that, Mama. You don't have that much savings and —"

The nurse comes in to remove your IV, cutting the conversation short. *Free at last, free at last!* You are glad to have that thing out of your arm. Wayne helps you put on your robe and slippers. It feels good to have on real shoes instead of those skid-free hospital socks. You ask Wayne to collect the shampoo and unused toiletries. You have a special drawer at home where you keep all the miniature supplies that you have collected over the years from hotels. Comes in real handy when you have company. Besides, it looks like you're paying for it anyway.

"Mama, we could be talking about *tens of thousands* of dollars."

"Oh dear." That gets your attention.

"I think we should at least *talk* to someone about this, Mama, I mean a lawyer. After what you just went through? I'll bet one good letter from a lawyer could persuade any drug

~ Page 359 ~

company to pay for your stay here. I mean, how does Crenshaw really know it wasn't the Kuritin? What do you think, Mama?

TURN TO PAGE 367 TO TALK TO AN ATTORNEY.

TURN TO PAGE 361 TO EXPLORE AN ALTERNATIVE.

# Gloria: Alternatives

YOU ARE GLAD TO BE OUT OF THE HOSPITAL. With each passing day you are feeling stronger. While still a little weak, you are able to fix yourself a piece of toast and cup of tea in the mornings. You enjoy sitting in the breakfast room and looking out onto your back yard. You love to watch the squirrels trying to steal seeds from the bird feeders. They always fall off, but dang, they never give up hope. *Bless their hearts.*

Walter's comments about Dr. Crenshaw's hiding some of the facts about the study bothers you more than your mounting medical bills. You wonder, *how much should trust cost? Have I been duped? Hell, if I can't trust my doctors anymore, in whom* can *I trust?* You start to think you should withdraw from not only the study, but medicine altogether. You don't even want to go the chemo/surgery route. *I wonder what they'll be hiding from me there?* You are getting sick of the whole healthcare system, literally.

As you contemplate the mess your life is in, you remember a business card your friend, Mirabelle, gave you just before she passed away. You search through the collection of torn papers with phone numbers, scraps of notes and business cards you keep in the little wicker basket on the kitchen table. You find the card: Henry Maze, MD, PhD, Traditional and Alternative Therapies. Mirabelle had told you that it was too late for her to try alternative medicine after she was diagnosed with cancer, but she wanted you to have it just in case you ever needed it. You figure "just in case" is here and now.

You call Dr. Maze's office and make an appointment. You feel comfortable about going, especially since he's a western doctor who incorporates eastern therapies whenever possible. You have always believed in natural cures. *A billion Chinese can't be all wrong, can they?* A sense of relief pours over you.

However, for this appointment with a doctor, you are going by yourself. You don't want to listen to Wayne telling you how crazy you are for visiting a "witch doctor," and, while you would love to tell Walter, you think it is best to leave him out of the loop for now.

The following week you take a cab to your appointment. You're feeling much better now that you are off that, "God-awful miracle drug," as Wayne calls it. Dr. Maze's office is small and inviting, not at all like the huge waiting room at the Crenshaw Center. A waterfall mounted on the wall makes soothing sounds. The other patients in the room look happy and hopeful — something you have not seen in a while.

You are greeted by the receptionist who asks you to fill out a questionnaire. So far there is nothing witch crafty about your visit. It seems like a regular appointment at a regular doctor's office.

You return the questionnaire and in a few moments are escorted by the receptionist to an exam room. "Dr. Maze will be in momentarily, Mrs. Brewster. Please make yourself comfortable. Would you like a cup of tea or some water?"

"No. I don't suppose you have coffee, do you?" You didn't see a fancy espresso machine in the waiting room.

"Sure we do. How do you take it?"

"A little cream."

"Is soy milk okay?

"I never tried it, but what the hey, I never tried this before either."

"I'll be right back."

You pick up a small brochure from a stack lying on a desk near the door. *Interesting stuff,* you think, although the terms, "bio-identical hormone replacement, customized vitamin supplementation, and treatment for yeast overgrowth," seem confusing.

Five minutes later, a middle-aged man with a receding hair line and pleasant round face enters the room. "Someone order a coffee?"

"Yes, thank you." You take the cup from his slender hands.

He sits on a stool across from you and adjusts the height so that his eyes and your eyes are at the same level. You are not separated by a huge coffee table. There are no credenzas filled with expensive museum pieces or portraits of the famous. It's a comfortable, cozy space, perfect for conversation.

"I'm Dr. Maze, but you can call me Henry."

"Pleased to meet you, Henry. I'm Gloria." You like this guy already.

"Tell me a bit about yourself, Gloria. I'd like to know a little about you before we talk about why you came in. Would that be all right?"

"Sure." You are surprised by his approach. He didn't start by taking your blood pressure like everyone else had. It's a refreshing change. You tell him how you grew up in a little Texas town and then went off to Austin for college. You tell him how you got married and had a son and how your husband was a sergeant in the army during the Vietnam War. You tell him a little of Wayne's domestic problems but mostly how proud you are of Wayne. You say that you have lots of friends and live in a good but old neighborhood. You tell him that one of your best friends came to see him a few years back and that's how you wound up here today. Finally, you tell him that you have a grandson in Austin going to medical school and that he is the light of your life.

Henry tells you that he too was raised in a small Texas town. He too was in Austin for medical school. After practicing family medicine for twenty years, he decided to go back to school to get a PhD in Asian studies. He chose Johns Hopkins University because of it's proximity to Georgetown University, where they had a master's degree program in Complementary / Alternative Medicine — something he had been interested in for many years. He learned to speak and read Mandarin, although he never mastered the art of writing Chinese. He has a wife and three kids, two in college.

You like his approach. It's a much nicer way of practicing medicine. You think you may be feeling better already. After about forty-five minutes he finally asks, "Well, okay, now. What can I do for you?"

You explain your story in excruciating detail. You are surprised as tears form in your eyes. You stop and pause to think of happier times before continuing, but nothing you can do holds back the tears.

Henry listens intently. When you are finished, he reaches toward a small desk and pulls open a drawer. Out comes a box of Kleenex. Henry hands you a tissue, saying, "I am so sorry that you have had such a hard time. Medicine today is not the way it used to be. That's partly why I stopped practicing conventional medicine. It was not supposed to be this way."

"You are the first person who took the time to listen to me, Henry. I can't tell you how good that makes me feel."

Henry looks at you carefully, then takes your hand. "Gloria, I wish I could help you, but for now I am not sure that I can offer a better solution than your other doctors. Most of our work has to do with prevention. There are some promising possibilities in alternative therapies for cancer treatment, but we don't have enough evidence to base any significant claim. I think it's too risky for you."

Your heart sinks, again, into your stomach. "I see."

"I'm not an oncologist mind you, but it's my understanding that Stage I oral cancer is eminently treatable with conventional therapies."

"I am not looking forward to that. Is there *anything* you can do, Henry?" There is more than a little desperation in your voice.

"If you decide to go the traditional route of chemo and surgery we can help manage pain with acupressure and acupuncture. We may also be able to add some supplements and work with you on diet, but that will depend on the chemo. You see, sometimes, vitamins and supplements can interfere with treatments, so we want to be careful about that. On the other hand, if you go back to the Kuritin study, I can't do anything."

"Well … I appreciate your time."

Dr. Maze stands up. "It was wonderful meeting you, Gloria. I wish you the best of luck. Please know that I will help you any way I can." He cups your hand in his. It is a warm, sincere shake. He looks you in the eye and you smile. You're sure Henry is a man who truly cares for his patients. Even if he can't help you with cancer, you are glad you came. You think that perhaps you might come back to him for some of his wellness therapies once you beat this disease — that is, if life doesn't beat you first.

TURN TO PAGE 357 TO TRY AGAIN.

# Gloria: Attorney

"THANK YOU FOR COMING MRS. PARKER." Wayne escorts an attractive young woman smartly dressed in a business suit from the front door to your living room sofa. Rachel Parker wears her hair in a tight bun and her wire-rim glasses on the end of her nose. She wheels a large leather box behind her and collapses the handle as she sits down. *Why that little girl's bag probably weighs as much as she does*, you think.

"It's my pleasure." Rachel passes out her business card.

"I hope you're going somewhere nice on vacation after this, young lady," you remark.

"Oh, this is not a suitcase, Mr. Brewster." Rachel chuckles. "This is my briefcase. This is your file."

"Call me Gloria. You say all that is mine?"

"Kind of. Some of it are your medical records. Then there are your bills and backups. I also asked for and received the KURE study enrollment forms and IRB submission. Lots of paperwork there. I threw in some case law too from LexisNexis. Anyway, I have a few questions I'd like to go over with you before I start my research."

"Knock, knock!" It's Walter. "Hey, Nani. Thought I'd stop by on the way to class."

"Wonderful! We have enough for a round of Uno! Walter, this is Ms. Parker. She's the attorney your father found to look into getting the drug company to pay my hospital bills. Believe it or not, Buddy, they're over $75,000!"

"Ouch," says Walter.

"Rachel, this is my grandson Walter. He's going to be a doctor. Walter, why don't you sit on the couch next to Rachel."

Rachel extends her hand.

"Good to meet you, Ms. Parker." Walter looks at you with a knowing smile as he shakes Rachel's hand, then steps over the coffee table and sits a few inches from Rachel. You wink at Walter *Well done, I must say. They really look sweet together.*

"Oh, please call me Rachel."

"Would you all like some tea?" *They'd make a charming couple, yes? She's probably not much older than Walter.* "Wayne, would you mind?"

"Not at all, Mama." Wayne heads toward the kitchen.

"Wow, Mrs. Brewster —"

"That's Gloria, Rachel."

"Sorry, Gloria. Look at all this art on your walls! They're fantastic."

"Yes, well, they go back many years. It's been some time since I've been able to hold a brush."

"You painted all of these?"

"A lifetime ago, yes."

"Wow."

"I'm glad you appreciate my painting, Rachel."

"I do. I do." Rachel opens her rolling briefcase and adjusts her glasses. "As I understand it, the doctors told you your medical emergency was not related to the study."

"*That's what he said,*" Wayne is shouting from the kitchen over banging tea pots and clinking china.

Walter's eyes widen. "Really? How could it not be?" Walter turns towards the kitchen door. "*Are you sure that's what they said, Dad? Are you talking about Dr. Crenshaw?*"

"*Yes, and yes.*"

Walter shakes his head. "I don't get it."

"Don't get what?" Rachel asks.

"Things aren't making sense." Walter pauses a moment. "I hate to say this … but I'm beginning to think something is not right here. I've been doing a little research on Nani's drug."

"Kuritin?" Rachel pulls a yellow legal pad from her case and starts writing.

"Yeah … I think Dr. Crenshaw was giving Nani too high a dose for her age. I found journal reports on Kuritin used in animal studies. The drug showed few side effects on young animals but all kinds of problems in older ones at the same dosage. Very similar side effects, in fact, as Nani's."

Rachel searches her case and finds a white folder marked with color tabs on the edge. "This is a copy of your chart, Gloria. I notice Dr. Crenshaw cut your Kuritin dose at your last visit to his clinic. Did he tell you why?"

"Yes," you reply. "He said it was because I was making such good progress."

"May I?" Walter takes the chart from Rachel's hands and starts flipping pages.

"Walter, I'd like copies of those articles you mentioned. Would that be okay? Here's my card. You can have them delivered or I can pick them up from you. Whichever is more convenient."

"Sure." Walter takes the card and studies Rachel's face. "So far as I can tell, this drug is not meant for older patients."

"That's my read too." Rachel pulls out another folder and places it on the coffee table. "According to the study protocol, no one over 55 years old should be taking Kuritin. I'm no physician, but I take that as meaning, at Nani's age, any amount of Kuritin could be risky."

Walter puts the chart down and picks up the protocol folder.

"*What?*" Wayne is standing with a tray of four teacups in the kitchen door. It takes a few seconds for things to sink in, but he soon gets mad. "What the *hell?*" Wayne walks into the living room. "Why would Crenshaw enroll Mom, then? Why would he put her at risk? Why? Does this have something to do with money? Is he making money from this … this *research?* You're telling me that *he knew!* That sonofa—"

"Calm down Dad. Nothing's for sure yet, and I don't think Ms. Parker —"

"Rachel."

"Rachel appreciates your swearing."

"I've heard much worse, Walter, but thank you."

Wayne is not calming down. "My God, he almost *killed* her! Worst of all, he had the nerve to stand there and tell me to my face that Mama's almost dying had *nothing* to do with his study! Then, to tell her to get back to taking his pills? Is he *nuts?*

"One step at a time, Mr. Brewster." Rachel's voice is soft but commanding. Wayne puts the tray on the coffee table and sits down. His head drifts into his hands.

"There's something else," Rachel says. She gently takes the protocol folder from Walter's hands and opens to a page bookmarked with a yellow Post-it note. "The protocol says all serious adverse events must be reported to the IRB within 10 days." Rachel takes off her glasses. "Inpatient hospitalization is considered a serious event. It's been two weeks since your emergency room visit, Gloria. I called the chair of the university's IRB this morning. Dr. Crenshaw has not reported a serious adverse event."

"What's an IRB?" you ask.

"What the *hell?*" Walter says.

Turn to page 372 to continue.

# Gloria: Investigation

TWO DAYS LATER, RACHEL IS BACK AT YOUR HOUSE, parked along the street. Wayne drives up a few minutes after Rachel arrives and parks in your driveway.

Rachel says the phone call from the university was cryptic. "All I know, Gloria, is that they want to meet with you about the Kuritin study. I assume they called me instead of you because they fear there may be a claim over your emergency room visit."

"Rachel," you say as Wayne opens her car's passenger door for you, "this is the first time in my life that I've ever been involved with the legal system."

Wayne closes the door then climbs into the back seat of Rachel's Honda Fit. "Wow. There's an amazing amount of room back here for a small car," he notices.

"Yeah, that's why I bought it." Rachel adjusts her rear view mirror. "Well, that and because it was cheap, but mostly because it's got room for my junk. It's sort of my office's file room." Rachel turns and helps you buckle up. "This is not a lawsuit yet, Gloria. I'm still doing research to see if we have a basis for filing a claim. My guess is that the university would just like to resolve this before things escalate."

You don't much like the idea of claims and lawsuits and such, but you don't like getting stuck with a huge hospital bill either, especially if it was because some bigwig decided to bend a few rules. Still, you have butterflies in your stomach as Rachel pulls her car under the administration building's porte cochere. It seems like Wayne is nervous too. He's barely said a word the entire car ride over.

Rachel drops you and Wayne off at the entrance, then takes a wheelchair out of her trunk before heading to the garage. She returns a few minutes later, pulling her briefcase behind her. "Twelve dollars. Can you believe it? Don't suppose they'll validate my parking, do you?"

You enter the lobby of an old federal style office building in a wheelchair pushed by your son. That was Rachel's suggestion. You are feeling much better and probably don't need the wheelchair, but everyone seemed to think it was a good idea, even Walter — who was probably more worried about you falling than the theatrics of it all.

Rachel walks to the information desk and is back in thirty seconds. "We're going to meet just down the hall."

You find a woman standing in an open door smack in the middle of a long corridor. She raises her hand as you approach, indicating that you've reached your destination. A bronze plaque on the door reads: "SAM HOUSTON BOARD ROOM." You enter, mildly impressed. The woman appears as stately as her surroundings, with gray braids that follow her hairline around the nape of her neck to form a tidy bun. "Dr. Cochran will be right in. May I get you all something to drink?"

"I'll have a coffee if you've got it," you say.

Rachel doesn't want anything. You figure she doesn't want to get too comfortable before putting her negotiating face on. Wayne is too self absorbed to hear the question.

"My son will have water if you don't mind, Ms., um ..."

"Bryan. Margaret Ann Bryan."

"I'm Gloria."

"Yes, I know. Gloria Brewster." Margaret Ann carries herself in a way that makes you wonder if she is a member of the Daughters of the Republic of Texas. Perhaps you'll ask her if the opportunity arises.

Wayne pushes you to a spot between two chairs at an ornate wooden table. It's very old and very large. You figure 30 people could sit around the thing and the table would still look half empty. The great seal of the university is inlaid with mother of pearl into the center of the table. The room smells like worn leather and old carpets. Forest green velour curtains hang from ceiling to floor on large brass rings. Plaster molding crowns the perimeter of the room. There is an intricately detailed wood baseboard. You can tell it was a grand room back in its day. Today, it seems like an heirloom.

"Gloria, you doing okay?" Rachel asks.

"Yeah. Fine."

Rachel sits in between you and Wayne. "Just remember to let me do the talking. I'll let you know when you can speak."

"That sounds good to me."

"Same goes for you too, Mr. Brewster."

"Yes, ma'am."

Margaret Ann returns with the drinks. She places a silver tray on the table and puts coasters emblazoned with the same university seal in front of you and Wayne. Shortly thereafter a gentleman and woman enter the room.

"Good Morning. I'm Dr. Cochran and this is Dr. Frey."

Rachel stands to shake their hands. "I'm Rachel Parker. This is Gloria Brewster and her son, Wayne." Rachel passes out her card as she watches Margaret Ann walk out of the conference room and close the door behind her. Rachel looks a bit surprised. "Will anyone else be joining us?"

"No," says Dr. Cochran.

"Are either of you an attorney?" Rachel asks.

"No."

"Then you are not represented here today by university counsel?"

"No." Cochran is a serious looking fellow, and he doesn't look particularly happy to be here answering Rachel's questions.

"You do understand that I *am* an attorney and that I represent Mrs. Brewster —"

Cochran cuts her off. "We got that Ms. Parker. You and I spoke on the phone, remember? This is not a legal issue yet. We asked you here today as part of the university's internal investigation. We just want to ask Mrs. Brewster a few questions about the Kuritin trial. That's all."

Rachel sits down as Cochran and Frey take seats directly across the table from your side. Cochran pours a glass of water from a pitcher sitting on the tray. He offers a glass to Frey, who declines. Cochran then explains that he is a chair of the university's Institutional Review Board. He introduces Dr. Frey as the university's Research Integrity Officer. Cochran explains how both of them are responsible for upholding the rights and welfare of research study participants.

"I see." Rachel pulls out a voice recorder from her case and places it on the table. "Would anyone mind if I taped our conversation for my notes?"

Cochran looks at Frey. "We'd prefer that you don't." Rachel puts the device back into her briefcase with an impish *can't blame me for trying* look on her face. *She really is two sweet to be an attorney*, you think. *She's not frightening at all.*

Cochran leans toward the center of the table. "Mrs. Brewster, I want you to know that we are sorry you have not been happy with your participation in the Kuritin study —"

"*Sorry? Happy?*" Wayne interrupts. "Hell! She's been at death's door."

Rachel puts a hand on Wayne's forearm, which stops him from going further. Looking directly at Cochran, she says, "Allow me to put this in simple terms, Dr. Cochran …" The sweetness in Rachel's expression evaporates. Her eyes widen as she pulls out a stack of papers and file folders from below the table. "We all know Dr. Crenshaw was not following his own study's protocol. As a result, my client became very ill and almost died. She also amassed over $375,000 in medical bills, twenty percent of which the university's hospital will be trying to collect from her." Rachel's voice is just shy of confrontational and getting everyone's attention. "That's $75,000 owed by my client."

Cochran's dour facial expression doesn't change. "The university will take care of Mrs. Brewster's medical expenses, Ms. Parker."

There are five seconds of silence before Wayne chimes back. "Damn straight."

"Mr. Brewster, please." Rachel's hand returns to Wayne's forearm, a bit firmer this time. Rachel studies Cochran's face. "*All* medical expenses? What about future treatments, Dr. Cochran?"

Now it's Dr. Frey. "All medical expenses, and we will continue to treat Mrs. Brewster for her oral cancer at no charge to her."

Another pregnant pause.

"Let me make sure that I understand this correctly, Dr. Frey." Rachel is writing on her yellow legal pad. "You are saying that the university will cover *all* of Mrs. Brewster's out-of-pocket medical expenses from her previous and future cancer treatments?"

"That is correct. We will follow her all the way through the course of her disease." Cochran drums his index finger on the table for emphasis.

"No matter what treatment she elects?" Rachel's index finger drums the table as well.

"Correct."

"You will put this in writing?"

"Yes."

Rachel puts her pen down. "Good, but one more thing, Dr. Cochran." Rachel does not skip a beat. "I am sure that you and Dr. Frey would agree that Mrs. Brewster has gone through a great ordeal. You may also know that she has fallen a number of times from weakness. She has been afraid to walk from her bed to the bathroom for fear of winding up on the floor and not being able to reach a phone for help. She has also lost a great deal of weight because she cannot keep most food down.

Cochran's expression does not change.

"The bottom line, Dr. Cochran, is that Mrs. Brewster has experienced an unnecessary but significant amount of pain and suffering as the direct result of what we believe was improper management of the Kuritin study. Besides covering all of Mrs. Brewster's medical expenses, we would ask for additional compensation for pain and suffering."

Cochran sits back in his chair. "Pain and suffering?"

"Yes."

"Well, … I …" Cochran doesn't like where this is going. He looks toward the end of the room at a credenza sitting against the wall. On the credenza is a telephone.

"I understand if you need to consult with the your legal affairs office. We can wait a short while before we file a formal complaint."

Now it's your turn to speak. In your best Texas drawl you say, "Rachel, I appreciate all that you're doing for me … but, this is unnecessary."

"I'm sorry?" Rachel peers over her glasses.

"Dr. Cochran and Dr. Frey said they would pay my medical bills," you tell her. "That's enough for me."

Rachel is silent for a moment, then turns to Wayne, who is looking intently at you. You are sure you can hear his thoughts. *Mama, you are giving up what could be a lot of money.*

You telepath back with your eyes. *This is not about money, son. This is about what's right.*

Wayne smiles. "She's the boss, Rachel."

Rachel pushes her glasses back on her face. "Yes, … she most definitely is."

At long last, Cochran changes his expression. He looks relieved, even friendly. "Thank you, Mrs. Brewster. Thank you for your understanding. Let me just say that, on behalf of the university, I apologize for what you've been through."

"Call me Gloria, Dr. Cochran. And didn't you say you had some questions for me about the Kuritin trial?

Turn to page 439 to continue.

# Parvesh: The Meeting

FOR SOME REASON you thought things moved more slowly in Texas. That there was a more lackadaisical approach to life. Well, it is certainly not so at this school, you think, as you read the email for perhaps the fifth time:

> Parvesh.
>
> Hello again, this is Dr. Patel. Do you remember speaking with me at Dr. Hernandez' party? I hope so, as things are about to move swiftly. I received this email recently:

There are symbols running down the left side of the next section, signifying Dr. Patel is quoting another email.

> < Dr. Patel:
>
> < Dr. Crenshaw would like to meet with
> < you next Monday in his office to
> < discuss taking over the research
> < coordinator role for our ongoing
> < clinical trial of the new drug, Kuritin.
> < As Dr. Crenshaw may have told you at
> < a recent function, our former
> < Coordinator left suddenly two weeks
> < ago, and he is very interested in the
> < possibility of you filling the role.
> <
> < Would a meeting at 9:30 AM next
> < Monday work with your schedule?

< Please let me know as soon as possible.
<
< - Joyce Whipley, Secretary for Lawrence
< Crenshaw, DDS, PhD, MD
< The Crenshaw Center For Oral Health

The quoted email ends, returning to Dr. Patel.

Recalling our conversation at the party, I asked if I could bring a dental student interested in research to the meeting, and here is what I got in reply:

< Dr. Patel: I relayed your question to Dr.
< Crenshaw, and he answered (and I
< quote), "Absolutely! We're in need of a
< young, enthusiastic research assistant as
< well!"
<
< - Joyce

And back to Dr. Patel:

If you were indeed serious in our conversation, please confirm with me. Also confirm with me if you are NOT interested in this opportunity so I can advise Dr. Crenshaw.

I must say this trial sounds very interesting, and if the drug is as good as Dr. Crenshaw is saying, it would not be a bad thing to have on your resume.

-Faiza Patel, PhD

You settle back in your chair. Yes, it does say what you think it did. No, you are not misinterpreting it. You, a first year dental student, are being offered a chance to assist in some cutting edge medical research. *What an opportunity!*

You hit "Reply" and accept. What were you supposed to be doing at 9:30 AM on Monday? Does it matter? At 9:30 AM next Monday, you are going to the Crenshaw Center.

TURN TO PAGE 384 TO CONTINUE.

# Parvesh: The Exam

THE WAYS IN WHICH WE MOVE and are moved through life are strange and wondrous, you believe. Here you stand in the lobby of the Crenshaw Center For Oral Health, iPad under one arm, feeling very professional indeed. Were you still in New York, you would be glowering at the grey skies of encroaching winter, cursing a climate that forces a man to armor up with multiple layers of clothing just to face the outdoors. Again, you offer a brief prayer of thanks for the University's letter of acceptance, allowing you to move to Austin, where the climate is more in keeping with that of your youth in New Delhi.

You tap on the iPad to familiarize yourself with the upcoming patient, or at least to make certain you are still familiar with the case. You may not have inherited your father's childlike delight in snow, but he did ingrain in you a respect for thoroughness, a discipline which played no small part, you are sure, in Dr. Crenshaw selecting you as his research associate in his latest study.

You look up at the portrait of the original Dr. Crenshaw, looking over the front door of the research center and clinic he founded. He was a handsome man. You could almost see in his smile that he favored his grandson, the current Dr. Crenshaw, greatly. As the current head of the Crenshaw Center, the latter day Crenshaw seems determined to carry on his grandfather's reputation as the country's primary figure in the fight against oral cancer. For the Crenshaws, it's something of a family business.

Your patient is one Gloria Brewster, 81 years of age, who presented at Dr. Nguyen's private dental practice with a suspicious lesion on her tongue. Dr. Nguyen's preliminary notes seem unconcerned, but her hygienist, Leticia, insisted Nguyen employ a Verascope. The results from that seemed ambiguous — and since none of your instructors can agree on the value of the Verascope anyway, you're not surprised that Nguyen decided a biopsy would be prudent. So, an appointment was made for Mrs. Brewster at the Crenshaw Center.

A man and a woman enter the lobby. The woman is obviously older than the man, but seems more spry. They speak to the receptionist, who points to you. You step forward to meet them, your hand outstretched. "Good morning!" you say, "I am Parvesh Singh. I presume you are Mrs. Brewster?"

The woman accepts your hand. "Please, call me Gloria."

The man offers his hand. "I'm Wayne. I'm the son."

"Ah, yes!" you reply, shaking his hand. "A pleasure to meet you both. Would you follow me, please?"

As the two follow you to the elevator, Gloria asks, "And what shall we call you? Dr. Singh?"

"Oh goodness, no," you reply, pressing the button. "Not yet, at any rate. Soon, it is to be hoped. As you are Gloria, I am simply Parvesh."

The three of you enter the elevator. "That's too bad," Gloria smiles. "Dr. Singh has a nice ring to it."

You laugh. "I cannot help but agree, but I would think you are talking to my father." You and Gloria exchange pleasantries all the way to the examination room. Wayne, on the other hand, remains silent and nervous. Gloria seems to be handling the possibility of her illness much better than her son.

In the examination room, you tell Gloria, "If you would change into one of those gowns, please? I will go see if Dr. Nguyen has arrived and we will try to get this done quickly."

"Quickly works for me," Gloria says. "My stories'll be on soon."

As you escort Dr. Nguyen to the examination room, you see Wayne Brewster standing outside the door. You are not surprised — you would have left the room to give your mother some privacy while changing as well, but as you arrive at the door, it becomes obvious that Mr. Brewster had other reasons to await your return.

"So, Dr. Nguyen," he says, "what is it? Is it serious?"

"It?" asks Dr. Nguyen.

"It. That thing on her tongue. Can you cure her? Is it even treatable?"

"Mr. Brewster …"

"And, and, and …" he says in a rush, "do you have any idea how much this is going to cost? I mean, do we even know if Medicare is going to have anything to do with this? I'm in a whole new world here!"

"Mr. Brewster," Nguyen says calmly, firmly. Wayne subsides at the sound of her voice. "I still don't know what *it* is. That's what this visit is about. We will address all your questions, believe me. Right now, we need to answer your first question."

"What *it* is," he replies softly.

"Yes."

"Right," Wayne sighs. "Sorry. I'm sorry. It's just …" His voice trails off.

You say softly, "We understand, Mr. Brewster. She is your mother."

"Yeah."

There is a brief silence as Nguyen reaches for the door. "Then let's get these questions answered."

You all enter the room. Gloria is seated on the exam chair. "I don't mind telling you, doctor, I had a conical biopsy for a cervical cancer scare a few years back, and I did not enjoy it one bit."

"I don't blame you," Nguyen responds. "I had one myself. No fun at all." She takes a tray of surgical instruments from a nearby cabinet, and sets it on a nearby table. "Unfortunately for us both," she says to Gloria, "I left my magic wand at a dental convention last month. Will have to use this instead." Nguyen gestures with a long swab, "I'm going to brush some anesthetic over your tongue."

Gloria dutifully sticks out her tongue and Nguyen swabs the area around the lesion. When she is finished, Gloria smacks her lips. "That tastes awful."

"That's odd. It's supposed to taste like piña colada,"

"Well, I hate piña coladas. That might explain it."

"How's your tongue feeling?"

"It's getting numb. I guess I'm going to be talking funny for a while."

"Gloria, I'm Vietnamese. You all talk funny to me." Gloria laughs.

You find it enlightening to watch Dr. Nguyen in action. Gloria seems to have lost any foreboding she may have had about the procedure, and even Wayne has lost some of his grimness and relaxed a bit.

After a few minutes, Nguyen feels the anesthetic has had time to do its work and moves the chair back until Gloria's head is lower than her feet. She dons a mask and gloves. She motions you to do the same so you can observe. Seeing Nguyen approach with some kind of shiny instrument, Gloria closes her eyes and dutifully opens her mouth. Nguyen adjusts the operatory light and speaks to her patient calmly. "Gloria, I need you to stay as still as possible. You might feel a little pinch, because I'm not telling you anything new when I say the tongue is a big mass of nerves and — oh, never mind, I'm finished."

Gloria opens her eyes in surprise, then squints in the light, which you rapidly swing away. "Reyee? I'm imprethed."

"As I said. You all talk funny." Nguyen works the foot pedals to return Gloria to a seated position. "Let me write up my notes so Parvesh can enter them into the Center's fancy new Electronic Health Record system, then I'm going to walk this sample up to the lab. If you come with me, I'll show you where you need to go for the blood work, and the CAT scan is close by."

Later, you transcribe Nguyen's notes, marveling that she seems to have avoided the curse of doctor's handwriting, and writes in a precise, uncluttered print. You check over your work and save it. Then you pull out your phone and tap Dr. Crenshaw's number.

"Crenshaw," comes the voice on the other end.

"Dr. Crenshaw, this is Parvesh. I may have another candidate for the study."

"That's good. Is the cancer confirmed?"

"Not yet. The biopsy was only taken this afternoon. I am a bit concerned, though, that the patient may be too old. She is 81 years old."

"That may not be significant. E-mail me her records."

"Ah ... yes, sir." There is a click. The conversation is apparently over.

You sigh and stare at the glossy face of your tablet computer for a moment. You feel bad for poor Wayne Brewster. Even the possibility of your mother developing something like cancer touches on a nameless dread which you would rather remained untouched.

You access the menu of the EHR program and touch the "Export/Email" button. A dialogue box pops into the middle of the screen. Alongside a large exclamation point inside a yellow triangle, it reads: "NOTICE: PROTECTED HEALTH INFORMATION (HIPAA PRIVACY RULE, 45 CFR 164.524) RULES MAY APPLY. PATIENT AUTHORIZATION MAY BE REQUIRED FOR EXPORT."

There are two buttons underneath the message. The green one says, "Continue." The one labeled, "Cancel" is the friendly color of fire engine red.

You blink. *Well, of course there is a message like this. But Crenshaw is the Director of the Center, yes? He must know what he is doing.*

TURN TO PAGE 390 TO TOUCH THE "CONTINUE" BUTTON.

TURN TO PAGE 392 TO TOUCH THE "CANCEL" BUTTON.

# Parvesh: Continue

THE TROUBLESOME EXCLAMATION POINT VANISHES, replaced by a drop-down list of data. You touch the words "Face Sheet," and they transform into a small picture of a folder under your finger. The folder slides itself over and disappears into the tablet's email icon. A small gray bar in the center of the screen quickly fills up with blue, then is replaced by the words "EXPORT COMPLETE."

A new email message opens with Gloria's record attached. You tap in three letters of Dr. Crenshaw's email address and auto-complete does the rest. You touch "Send," and that is that.

*Now, my friend,* you think, *it is time for lunch.*

TURN TO PAGE 391 TO CONTINUE.

# Parvesh: Alert

YOU HEARD THE PHRASE "MAD WHIRL OF LIFE" ONCE, and at the time, you thought you knew what it meant. *Well,* you think, *such is the ignorance of youth.* A young boy who never had to wolf down (*and why do Americans call eating, "wolfing" anyway?*) lunch sitting at a lab table in one of the world's foremost oral cancer centers. You have found that appending, "One of the world's foremost oral cancer centers" to many of your activities makes you feel better about your current lack of an individual life.

Your smart phone, sitting on the lab table nearby, buzzes angrily. *Does the poor thing feel ignored?* Lately, the only time you have had to check email is between bites at lunch. Well, to be sure, you are still chewing, but perhaps today we can cheat a bit.

The screen comes to life. You have perhaps twenty new emails — not unusual — but the most recent one catches your eye: The return address read "Medical Records" and the subject line is simply, "ALERT." You tap on it. *Somebody has not discovered it is possible to release the Caps Lock key?*

"THIS OFFICE HAS DETECTED WHAT MAY BE AN UNAUTHORIZED TRANSMISSION OF PROTECTED HEALTH INFORMATION FROM YOUR EMAIL ADDRESS. PLEASE CONTACT ..."

You realize you have stopped chewing because your lunch has turned to dull cardboard in your mouth.

TURN TO PAGE 384 TO TRY AGAIN.

# Parvesh: Cancel

YOU STARE AT THE DIALOGUE BOX, HESITATING. Perhaps it is that huge exclamation mark, but something about the situation bothers you. You touch "Cancel" and shuffle through Gloria Brewster's paperwork, looking for a form that would authorize using her personal health information for research purposes. Alas, you do not find one. Almost reluctantly, you close the file containing her EHR.

*Hmm. How am I going to explain this to Dr. Crenshaw?*

TURN TO PAGE 393 TO CONTINUE.

# Parvesh: Test Results

A WEEK LATER, YOU ARE AGAIN escorting Wayne and Gloria Brewster to an exam room. The intervening week has not been kind to them. Wayne is looking particularly haggard, and Gloria is not the same easygoing woman you escorted earlier. Perhaps uncertainty is taking its toll, perhaps there are other forces at work.

After a brief wait in the exam room, Dr. Nguyen enters, accompanied by Dr. Crenshaw. Wayne looks at the two doctors and points to the exam room door. "Can we...?" The doctors nod and step back out into the hall, followed by Wayne and yourself.

Wayne looks at Crenshaw and Nguyen. "There's more than one of you. So I guess ..."

"This is Dr. Crenshaw, Director of the Crenshaw Center," Dr. Nguyen says. "I'm sorry, Wayne, but the results are positive. It *is* cancer. Squamous cell carcinoma, Stage I."

Wayne's expression does not change — he always seems to look in pain — but the color drains from his face. He leans against the wall for support. "Stage I?" he asks. "How bad is that?"

"It means it hasn't spread yet but —"

Nguyen is interrupted by an insistent chirping from her jacket pocket. She takes a smart phone from her pocket, glances at it, and mutters something sharply in what you presume to be Vietnamese. "This is an emergency, I've got to go." She places a hand on Wayne's arm. "Wayne, I'm leaving you with Dr. Crenshaw here. You are in the best of hands. Don't worry, we'll get through this."

Crenshaw extends his hand. Wayne accepts and weakly shakes it, still trying to comprehend what he has been told. "Mr. Brewster, I'm terribly sorry that the news couldn't have been better."

Wayne looks at Crenshaw as if it is the first time he's seen him. "You the fella in that picture at the front desk?"

Crenshaw smiles. "Not quite, but I might as well be. The good thing, Mr. Brewster, is that oral cancer is what we do. In the next couple of days we're going to be talking to you about treatment options, but right now I'm going to ask my associate here, Parvesh, to fill you in on something that might help."

You feel an unaccustomed warmth as Dr. Crenshaw refers to you as "my associate," a bit of pride, but you must concentrate on Wayne. "Mr. Brewster, our center is performing a clinical trial of a new anticancer drug the company is calling Kuritin. It is designed to retard and hopefully prevent the spread of cancer cells, working on a genetic level to improve the body's chances of complete remission."

"It sounds almost like science fiction," says Wayne.

Crenshaw says, "Not science fiction, not anymore. In worldly terms, this drug is a rock star."

"So, how's the trial been going so far, Parvesh?"

"Results have been encouraging, but —"

Crenshaw interrupts, "Parvesh is being modest. Results have been incredible, and I think when we're finally ready to publish, we're going to be playing a whole new game, treatment-wise."

"Wow," says Wayne, then another thought occurs. "How much is something like that going to cost? I mean, if it's experimental, is her Medicare going to cover it?"

"That's the best part," answers Crenshaw. "The Center and the pharmaceutical company behind the trial are going to be eating the costs. Since it's integral to the study, quite a bit of Gloria's treatment will be free."

It is obvious that some weight has fallen from Wayne's shoulders. "Well, that's … that's great."

Now there's a beep sounds from Crenshaw's pocket. Sounds like an alarm reminder. "I have a meeting," he tells Wayne. "Parvesh, you have the paperwork, correct?"

"Yes, sir. Right here."

"Good. Wayne, I do believe in this drug. I think it represents your mother's absolute best chance for survival. I hope she'll join the study."

Wayne thanks Crenshaw and shakes his hand again before the man leaves. Then he turns to you and says, "Let me do the talking, okay? Just saying the word cancer would probably set her off."

You think Wayne is misjudging his mother's resilience, but you nod and usher Wayne through the exam room door. He approaches his mother, saying, "Good news! The docs say there's nothing to be worried about. They've just got to figure out the treatment and then you'll be cheating all your girlfriends at Uno again."

Gloria takes a deep breath, then lets it out. She seems to grow younger by ten years. "Oh, thank God." She stares at her son. "I *do not cheat* at Uno, Wayne. I *win*, fair and square."

You begin shuffling the various consent forms for the trial so that Gloria can more easily go through them and sign. You tell her, "We have a few forms for you to go over and sign, Mrs. Brewster."

"Parvesh, I left my reading glasses at home, dang it, otherwise I'd have been reading while y'all were palavering out in the hall. I guess you could read 'em to me, and I could sign them."

"Certainly," you say, preparing to sit next to her. "This first form is —"

Suddenly, Wayne is standing before you, pulling the forms from your hand. "Tell you what, Parvesh, you're a busy guy. I'll take care of that and we'll get these papers signed. Can I just drop 'em off with the girl out front?"

You look at Wayne uncomfortably. "I suppose so ... but if there are questions —"

"We'll flag them and ask before signing. Is that okay?"

"Well, ..."

Wayne gently pushes you to the door. "Thank you for everything, Parvesh."

As he closes the door behind you, you hear Mrs. Brewster exclaim, "Dr. Crenshaw? The fella in that painting up front?"

You ponder a moment before you head down the hall. Wayne's taking over of the informed consent process has given you some extra time before your next class, but that episode back there is bothering you. Mrs. Brewster makes the last of the new recruits Crenshaw had tasked you with finding, but the amount of duplicity Wayne is demonstrating ... *how long can he possibly hide from her the fact that she has cancer?* You think that this is how things may work in India, *but here?* How is he going to reconcile this *there's nothing to worry about* with cancer? You sigh. That, you try to tell yourself, is not your problem. Then you consider that perhaps it should be your problem, and your head begins to ache. Perhaps you should go back to the exam room and see how they're doing?

TURN TO PAGE 400 TO RETURN TO THE EXAM ROOM AND SAY SOMETHING.

TURN TO PAGE 398 TO STAY SILENT.

# Parvesh: Stay Silent

WAYNE'S DISMISSAL OF YOU STILL RANKLES A BIT, but at least it gives you an opportunity for a more leisurely lunch. Then there is little time for second thoughts, as your afternoon classes demand your full attention.

5:00 PM - You are home, such as it is: a small apartment that would make your mother sigh in frustration, but it suits your needs. A Bhangra music station playing over the web provides background noise as you begin your studies.

9:00 PM - Goodness, you certainly are hungry all of a sudden. Small wonder — you had an early lunch, thanks to Wayne's takeover of the release forms. You mull that incident over as the microwave heats your dinner unevenly. Then, back to the books.

2:00 AM - You have read this paragraph at least three times, and you still have no idea what it is about. Perhaps the tendency of your eyes to stay shut when you blink is a clue? Time for bed.

4:00 AM - *You knew the White Coat Ceremony was going to be a big deal, but the number of students ahead of you in the line is beyond ridiculous. Don't these people know you are due in class any minute? You could swear you see some faculty members in line, too. Don't they already have white coats?*

*Much sooner than you expected, you are at the head of the line. The student ahead of you accepts her white coat. The crowd cheers, and you hear the choral portion of Beethoven's Ninth Symphony. She glides off, and you find yourself standing before the Dean.*

*The Dean looks down upon you, as if from a great height. With a gesture of disgust, he turns his back to you. You feel an insistent tug on your arm, and discover it is the Assistant Dean, pulling you out of line.*

*"That's not for you, Parvesh," she is saying. You protest and glance back at the line, only to discover that the Dean has become an enormous Naga, his snakelike lower body scattering and smashing the students who had been behind you in the line. You hadn't noticed the flames surrounding you while you were standing in line, nor heard the fire alarm.*

*The Assistant Dean continues, "Do you remember Gloria Brewster?"*

*Speechless, you nod. The Assistant Dean's head is weaving back and forth as her lower body becomes that of a snake. You desperately wish to run, but cannot.*

*"Good, that's good," the Assistant Dean croons. "We need to have a little chat about that ..." and she smiles, revealing her fangs.*

*The fire alarm grows louder,* and you open your eyes, your left hand searching groggily for the alarm clock. You sit in the bed for a long while, trying to purge the imagery of the dream from your mind.

*Oh my goodness ... my subconscious can be a real jerk sometimes.*

TURN TO PAGE 393 TO TRY AGAIN.

# Parvesh: Say Something

YOU BEGIN TO WALK BACK TO THE LAB. This would be a good opportunity to take a leisurely lunch for a change, perhaps taking time to actually taste your food. Normally such a rueful observation would make you smile, but not this time. You decide that the situation in the exam room *should* be *your* problem, and you gracefully pirouette 180 degrees — polished terrazzo floors are a wonderful invention — and head back to the exam room … just in time to meet Gloria and Wayne as they exit.

Wayne shoves a stack of papers into your hands. "Here you go, Parvesh, all signed and ready to go! See ya!" Gloria waves goodbye and seems to want to stay and chat, but Wayne is hustling her toward the exit.

You look at the forms in your hand. Yes, they all seem to be signed. You exhale in resignation. Your intentions were good, which, according to your father, would make them excellent paving stones on the road to Hell.

TURN TO PAGE 401 TO CONTINUE.

# Parvesh: The Lab

THAT EVENING, YOU FIND YOURSELF in Dr. Crenshaw's "working" office, surrounded by rich wooden paneling you suspect has gone unchanged since the original Crenshaw built the research center and clinic after the Second World War; simply additional layers of oil applied. It is just one more way the spirit of the Center's founder seems to permeate everything done here.

The lab office is a miniature version of Crenshaw's larger "formal" office near the main lobby. Other than size and splendor, the only other difference between the two are the towers of paper, some six feet tall, stacked around the room. Seated with you before Crenshaw's antique desk is Dr. Patel, who, after your chance meeting at Dr. Hernandez' party at the beginning of the year, was instrumental in you obtaining your current position as research assistant. As Crenshaw looks through the printouts she brought to the meeting, continuously making disapproving sounds, Patel seems to perhaps regret her involvement.

"I don't know …" Crenshaw is saying. "This just seems … incomplete."

"Yes, I know it is incomplete," Patel replies. "I'm still digging through the previous study coordinator's data. By the look of things, she left rather abruptly. I can't figure out why —"

"No, no. I mean there just seems to be holes in it."

"I can assure you, I am being very thorough."

Crenshaw makes an unconvinced noise as he flips through the printout. "Tell you what, Faiza. Send me your master spreadsheet and let me look at the data. If there are any holes, I'll fill them."

"There aren't."

"Well, then, I won't have to do anything to them, will I? I'll rest easier. At your earliest convenience please." Then he turns to you. "You and I, Parvesh, have a completely different problem."

Apprehensively you ask, "Yes?"

"Recruitment for the study has got to be stepped up. If we don't find another five subjects in the next 60 days, the pharmaceutical company is going to cut us out of the trial. I can't let that happen."

"I thought we had reached our recruitment goal?"

"It's been raised. The drug company increased funding for the trial. I thought I told you."

He didn't, but never mind. "How should I go about this, sir?"

"You've got to be creative, my boy. The Brewster woman today, now there's an excellent example. Keep thinking outside the box, Parvesh. Don't let yourself get hemmed in by conventional thinking. Consider more than just those you think are ideal subjects."

"I thought her age might be a concern."

Crenshaw waves away your worry. "Let me make judgment calls like that. Don't let nitpicking details limit you. Our center has swarms of people passing through. Start going through last month's EHRs and see if anything jumps out."

Perhaps you should not have been joking about a degree in bioinformatics at Dr. Hernandez' party. Looks like fate has conspired to make you spend much of your day immersed in Electronic Health Records.

Crenshaw leans back in his chair, its antique leather creaking comfortably. "Look, there is significant money and a hefty portion of my reputation riding on this study. We have got to make our work *sing*, if you'll pardon the expression, Mr. Singh."

"Yes, sir," you say, the problem turning over and over in your mind like a wooden puzzle with missing pieces. There are going to be mountains of data to comb through. *What was the name of that redheaded lady at the party? The one who with a bioinformatics degree?* She might know how to speed up a search like this.

Crenshaw dismisses you and Patel from the meeting, but you hardly hear him. Your father grew fond of the Americanism, "Work smarter, not harder," and you are entirely preoccupied with how you are going to accomplish that.

TURN TO PAGE 404 TO CONTINUE.

# Parvesh: First Month

YOU ARE BACK IN AN EXAMINATION ROOM with Wayne and Gloria Brewster. Wayne is looking particularly haggard this morning. You begin to wonder if he himself has some underlying health problem. Gloria is keeping up a brave front, joking with Dr. Crenshaw as he goes over the drug protocols, but there is a stoic undercurrent to her bravery — Gloria seems to be bracing herself for the worst.

You take her blood pressure and note it. She does not complain as you draw a blood sample and mark the vial.

Dr. Crenshaw hands her a card of blister-packed capsules. "This is the miracle we're working on," he tells her. "Take one of these every day after breakfast. Then we'll see you again in a month and find out what sort of progress you're making."

Gloria takes the card. "Your lips to God's ears." She looks at the milky white capsules. "Not real purdy, are they? They look like the worm medicine I used to give my dog."

Crenshaw shrugs. "What can I say? Finding the most attractive colors for the capsules is another research group entirely."

"Doctor, I'll give you a pass this time. Should I take one now, or …?"

"Tomorrow. Right after breakfast."

"I can do that." She smiles at you. "Parvesh, you need to poke me anymore?"

You return the smile. "Not today, ma'am. I think we've taken quite enough from you."

"Okay then," she says, and allows Wayne to help her down from the table. "I hope to see you gentlemen in a month." As she passes you, she leans in slightly, whispering, "And it's Gloria."

"Yes, ma'am, um, Gloria." You smile and are rewarded with an amused, "Oh, you …" as mother and son exit. You quite like Mrs. Brewster. You truly hope this drug is everything Dr. Crenshaw claims.

\* \* \* \*

Three days later, you are in the lab preparing for your morning's duties when your phone chirps. The caller ID identifies it as Wayne Brewster, and immediately you know it is going to be interesting.

"Parvesh, we got a problem," comes a frantic voice from the phone. "My mom's been up all night vomiting. She's better now, but still rocky."

"My goodness," you say. "She is not keeping anything down?"

"Well, right now we're not having success with 7-Up and Saltines. That's what she always gave me when I was sick. I don't know what else to do." There is a voice in the background, "She says she'd like to try some chicken broth or something, but I'm kinda freaked out here, Parvesh. Should she even try to take today's pill?"

*This is going to require more than a first year dental student's expertise.* "Mr. Brewster, let me consult with Dr. Crenshaw. His office is just down the hall. I will have him call you immediately. Is this a good number for that?"

"Yes, Parvesh, I'll be staying right here. Thank you."

You knock on Dr. Crenshaw's heavy oaken door and are rewarded with a call of "In!" You enter and explain Mrs. Brewster's situation to Dr. Crenshaw. He nods and lifts the receiver of his phone. You read off Wayne's number to him.

"Mr. Brewster? This is Dr. Crenshaw. Tell me what's going on." The doctor is silent for a few moments, except for an occasional "uh, huh" and nod of the head. Finally it seems Wayne has run out of steam, and the doctor says, "I'm sorry to hear she's having this much trouble. I've got to tell you, Wayne, this is a powerful drug we're working with here. Initial reactions like this are common. Her body is undergoing a period of adjustment and you shouldn't be worried about it. Keep her on the regimen." He listens for a moment. "Yes, chicken soup sounds like a good idea, then the pill. Oh, no, no, no, it's not a bother. Please call if there are more developments. You're welcome. Have a good day."

Crenshaw hangs up his phone, looking satisfied. "You see, Parvesh? No problema." There is a brief pause as he waits for your response. "That means we don't have a problem."

"Yes, sir. Thank you, sir. I'll get back to the lab."

"Good man," says Crenshaw, already flipping through folders on his desk.

On the walk back to the lab, you keep considering Crenshaw's statement, *reactions like this are common.* Something about that is bothersome. Maybe the effortless way he glossed over Wayne's concern? It is puzzling.

In the lab, Faiza Patel is entering trial data into her computer, fingers flying over the keyboard. You envy her speed on the keyboard. You should have practiced touch-typing more in school. She glances up. "Parvesh, good morning," she says brightly. You wave distractedly at her and sit down in your work area.

Patel has now stopped her tapping at the computer and is considering you. "If you don't mind my saying so, you look like someone has stolen your lunch money."

You look at her, puzzled. She shrugs. "I saw it on TV last night. What is troubling you?"

You tell her about Gloria's reaction to the drug. Patel nods and flips through a pile of forms nearby, pulling out one with enviable certainty. "Here on the informed consent it says, common side effects include loss of appetite, nausea, and vomiting."

You feel better, until you realize that it is also possible that Wayne never read that part to her.

Patel, unmindful of your misgiving on that point, has returned to her computer. "Let's see how many other participants have reported this." A few keystrokes, and she says, "Oh, that's very interesting."

You rise and stand behind her for a better view of the screen. "Yes?"

"Only two." A few more keystrokes. "Overall, that's less than 3%." She turns to look at you again. "Can that possibly be right?"

You spread your hands to show you have no idea. Patel has turned back to the computer, minimizing her spreadsheet and bringing up her browser, pointing it to PubMed. Her fingers are dancing over the keyboard again. You watch, hoping to pick up some pointers. This woman is a warrior when it comes to research.

After a few failed queries, she sighs. "Unfortunately, this protocol is too new to have much of a literature. Here … there are two articles about animal research studies on the same drug and … oh, but of course, we do not subscribe to the electronic versions of those journals."

"What about the university's medical library? They may have printed versions available," you offer, glad to at last be able to contribute.

"That is an excellent idea," smiles Patel. She dances over to the university medical library website. "Look, you are correct. They have both journals in the stacks."

"Very good. I have no participants coming in for another hour or so. Shall I go over and get copies of the articles?"

"That would be wonderful. Here you go." More keyboard flourishes and mouse clicks. "I'm printing out the journal names, issues, and page numbers."

"Thank you. I'll be back in a bit."

"No, *thank you*," says Patel. "Libraries don't agree with me. I am allergic to dust." Then she is back to the computer, fingers tapping away like machine gun fire.

TURN TO PAGE 409 TO CONTINUE.

# Parvesh: The Library

DR. PATEL'S FEARS about the dust in this library are unfounded, you think as you weave through the stacks of the university's publication collections. *I wish the cafeteria were this clean.* Patel's printout in hand, you scan the bound journals' spines. These in particular date all the way back to 1910, impressive to say the least, but the one you are looking for is not there.

You move to the neighboring line of shelves, passing several carrels on the way, most packed with students and piles of books. Suspicious snores emanate from more than one of the cubicles.

The next line of volumes brings you more luck. Your finger follows a bumpy track down the line of books until you find the bound copy you need. The next shelf yields the second journal you require, and now you must search out the nearest copy machine.

As you swipe your student ID card through the machine's card reader and engage in a ballet of pressing the journal's pages against the glass, flipping to the next page, pressing again, you feel rather clumsy and primitive next to Patel's effortless juggling of windows on her monitor. Perhaps that is the reason she avoids the library. You do rather feel like you've reverted to frontier medicine here.

You staple together your copied pages and begin to stroll back to the elevator. You flip through the pages as you do, occasionally altering your course, as you sense another student

— sometimes equally absorbed in paper — on a collision course.

The first abstract reports some common side effects, such as fatigue, skin rash, diarrhea, loss of appetite, nausea, vomiting, difficulty breathing, cough and infection in greater than 20% of the test animals. *Greater than 20% develop nausea*, you think. *That is much more than 3%.*

The second abstract goes into greater detail, breaking down the test animals into age divisions. The younger subjects display the same side effects, but older animals developed additional symptoms: itching, dry skin, abdominal pain, liver damage, eye problems, lung scarring and inflammation, skin blistering, liver failure, sores and bleeding in the digestive tract, liver and kidney failure resulting in death, and hearing loss.

You stare at the phrase, *liver and kidney failure resulting in death* as you press the down button. That wasn't on the form Dr. Patel read to you, was it?

Two more students get into the elevator behind you. One is speaking into his cell phone *sotto voce*, but you are still able to hear him in the elevator. "Yeah, Dad, definitely keep pressing the fluids on her. Vomiting is very dehydrating, and if we don't get that taken care of, things are just gonna get worse. I'll check back with you after class and see how Grandma's doing. Maybe I can even dig up some stuff on this Kuritin she's taking."

The name Kuritin makes you glance up from the reports. You recognize the fellow on the phone from the same party where you met Dr. Patel — *oh, right, this is the fellow who ran his car into a tree. What was his name? Oh yeah, Walter. Well, he appears to be doing fine now.*

A series of switches seem to click together in your head. Gloria Brewster has mentioned "my grandson, Walter, the medical student" several times ... and ... *Oh my goodness! It is indeed such a small world. Or ... maybe ... I mean ... Has Bhagya brought us together? Should I tell Walter about these articles?*

TURN TO PAGE 413 TO TALK TO WALTER.

TURN TO PAGE 412 TO CONTINUE WITHOUT TALKING TO WALTER.

# Parvesh: Ignore Walter

YOU CONSIDER SAYING SOMETHING TO WALTER, but you are at a loss. *What can I say? I'm not sure I know anything yet.* In any case, you consider discussing a patient's medical status outside a clinical setting could be … problematic?

The elevator reaches the ground floor, and you both go your separate ways.

TURN TO PAGE 415 TO CONTINUE.

# Parvesh: Talk to Walter

THE ELEVATOR'S DOORS OPEN and its occupants resume their walks toward separate fates. You decide that, perhaps, it was meant that you should speak up. "Walter, isn't it?"

Walter turns around, slightly surprised. He looks at you, puzzled. You gesture to yourself by way of explanation. "Parvesh," you remind him.

A look of recognition dawns on his face. "That's right, Dr. Hernandez' party. I think I have a couple of pictures of you."

"I believe so. I could not help overhearing …"

Walter glances down at his phone. "Oh, sorry."

"No, no, no," you interrupt. "Gloria Brewster is your grandmother?"

"Yeah?" asks Walter, the puzzlement returning to his face.

"I am a research assistant in the clinical drug trial in which she is participating."

Again, you see comprehension transform Walter's face. His eyes widen. The elevator doors reopen, unleashing a flood of students. Walter beckons to a less crowded section of the library's lobby. "Let's take this over there."

Once you have found some breathing room, Walter asks, "What can you tell me about this Kuritin?"

You shrug. "Not much, truthfully. Our data analysis is barely underway. However, Gloria — your grandmother — is one of the few who has reported nausea."

"Few?"

"Less than three percent according to our data."

"Hmm. The way that was mentioned in the release forms, you'd think it would be more common."

You feel a sense of relief, guilty though it may be, wash over you. *At least someone in the family bothered to read the consent forms!* "Yes," you agree, "and that is one of the reasons I am here, looking for papers on the early stages of the drug's testing."

Walter is obviously thinking. "Looks like you beat me to the punch. I was going to start researching it too; try to figure out what we're dealing with here."

You think a moment, then pull Patel's printout from the papers you are carrying, the sheet bearing journal names, dates and page numbers. You no longer need it. "Here," you say. "It is a starting point, anyway."

Walter looks at the printout. "Thanks. I wasn't sure when I'd get time to do it, but this'll help."

"Right then. I must get back to the lab. Good luck."

"You too," says Walter, looking at the printout as if it was a treasure map. "You too."

TURN TO PAGE 415 TO CONTINUE.

# Parvesh: Second Month

IT IS TIME FOR GLORIA'S NEXT FOLLOWUP at the Crenshaw Center, and you are disheartened by her appearance. At her last visit, she walked through the lobby and into the examination room by herself. Today she is supported by Walter, and moving slowly.

As Crenshaw examines her, you take quick notes. She feels weak and her legs and feet are swollen. She is still experiencing nausea, and she adds, "I gotta tell you, I am having a hell of a time peeing, pardon my French, Parvesh. That's not a problem I've had since giving birth."

Walter speaks up. "That's worrying me, Dr. Crenshaw. Wasn't there a record of kidney failure in the early trials?"

Crenshaw shakes his head. "In the early trials, yes, and that's why animal experimentation is so important. Of course, we're dealing with a different version of the drug here."

"Doc," Gloria says, "I don't know how much more of this I can take. I've had friends go through chemo and this seems just as bad, if not worse. I'm not sure I can stay in the trial."

You've been working with Crenshaw for a while now, and you can see things that others likely cannot. There is a brief flash of irritation behind his eyes, perhaps, but then it is gone and the smooth Crenshaw delivery returns.

"That's probably the most common thing we hear in the cancer field — that the cure is worse than the disease. I know this sounds backwards, Gloria, but the symptoms you're experiencing are proof that the drug is working. You may not feel like it right now, but you are making incredible progress with the Kuritin. So much so that I feel safe reducing your dosage, and that should increase your comfort level as we continue your treatment. How about it, Gloria? Can you give me another month?"

Gloria sighs. You can tell she wants to believe. "I'm making good progress, huh?"

"Absolutely."

"Well, I guess I'll keep going. I hope I start feeling this 'progress' you're talking about soon."

"Great. Parvesh will finish up here while I go arrange for the reduced dosage."

Then Crenshaw is gone, leaving you with Gloria and Walter. Both are lost in thought, which you reflect is a good thing. You don't feel much like talking, either.

TURN TO PAGE 417 TO CONTINUE.

# Parvesh: Adverse Event

YOU HAVE HEARD THE PHRASE, "COMFORT FOOD" many times before, but never truly understood it. Not until now, as you open the package that came in the mail today and pull out several bottles of your mother's very own curry mix. You have tried to acclimate yourself to your current home, but as much as you've come to like this "Tex-Mex" to which all your friends seem addicted (and the Tabasco sauce *is* tasty), nothing is going to taste as good, or be as soothing to your tired, exhausted soul as the curry you are going to make tonight.

So, of course, your cell phone rings. You recognize the displayed caller's name. "Hello, Walter," you answer a bit too brightly, anticipating your curry.

"Parvesh, it's Walter," comes the voice on the phone, too anxious to have heard your greeting. "I just found my grandmother passed out in the living room. I've gotten her to the bed, but she's been vomiting again and I don't like the way her breath is rattling."

"Oh my goodness. I will call Dr. Crenshaw immediately. May I give him your number?"

"Please."

You phone Dr. Crenshaw and alert him to the situation. As he hangs up, you realize that you are still gripping one of the bottles of curry powder, perhaps a bit too tightly. *This is terrible,* you think. *Gloria has been through so much with this trial. Is it worth it?* It dawns on you that you have started thinking of her as

Gloria, not "Mrs. Brewster," and you are a bit surprised how happy that makes you.

You organize the necessary ingredients in your tiny kitchen to make the chicken curry. You have no *naan*, but plenty of jasmine rice, and just the anticipation of the smells of cooking has your mouth watering. *Now, as my mother always told me, first we wash —*

The cell phone rings again.

*This is what it is like to be a medical professional, I am sure.* You thumb the answer button. "Hello?"

"It's Walter again, Parvesh. My grandmother's gone into convulsions."

"*What?* Oh, no! Have you called an ambulance?"

"Yes, I called them before I called you the first time. The EMTs are already here. We're taking her to the emergency room. I need you to call Crenshaw again. He only shows up as Unknown Caller on my phone. I need somebody to tell the doctor on call about the drug she's on."

"Yes, of course. Which hospital?"

You jot down the name and immediately call Crenshaw. His tone seems — what? *Annoyed? Puzzled?* In any case he tells you he will take care of it and rings off.

You look at the ingredients for your dinner, crowded but neatly arranged on the kitchen counter. Your mouth is no longer watering in anticipation. You wrap the uncooked chicken in plastic and, with a sigh, replace it in the refrigerator. Then you retrieve your car keys and set out for the hospital.

✳ ✳ ✳ ✳

Even on a weekday, the waiting room for the ER is heavily populated. No one even looks up as you enter and present yourself at the front desk. The receptionist seems a bit mystified that someone so — shall we say, East Asian — should be asking for Mrs. Brewster. After a quick explanation, you are directed to Exam Room 4, down the hall and through two sets of swinging doors.

You find yourself in another brightly lit hall, either side lined with huge panes of glass. The sliding doors to each room are mostly open, with curtains providing privacy to those within. You scan the numbered signs at each doorway, searching for number four.

You have just found it when you hear a female voice behind you. "Parvesh?"

You turn to find someone standing behind you, looking very much at home in her scrubs. There is a smile playing across her dark face, which is maddeningly familiar.

"It's Cheryl. Remember me?"

"Oh, yes! Dr. Hernandez' party. I didn't recognize you with your clothes on … I mean without your bathing suit … Oh, I'm sorry. I mean …"

Her smile turns into a smirk. "You *are* a devil with the women, aren't you?"

You try to recover. "You must be on your …" unable to think of the word, your index finger makes circles in the air.

"Rotations, yes. I'm in the ER this week. *Say!* Did you ever score a date with that cute redhead you were ogling at the party?"

"Googling a redhead? Oh, goodness, no!" You feel yourself blushing. "Not after you told me that she is the Research Integrity Officer!"

Cheryl shakes her head. "You and John are the same, such cowards. All talk." Cheryl's smile turns to a quizzical look. "So what brings you down here?"

You quickly explain to her about Gloria, the drug trial, and your position as research assistant.

"Crenshaw, huh?" Cheryl says. "You *are* coming up in the world. I just saw him by the nurses' station consulting with Dr. Vu."

You glance down the hall and see Crenshaw at the end. He's talking to an Asian man in green scrubs, who is nodding and making notes on a tablet computer.

"She's in rough shape," Cheryl continues. "The EMTs said it was touch and go with her on the way here, but she appears to be stable now. We're about to move her to the ICU."

There is a metallic crash from one of the exam rooms and Cheryl propels herself toward it. "Duty calls! Check you later, Parvesh!"

Down the hall, you see Dr. Vu also trotting toward the sound, his face a mask of concern. Dr. Crenshaw, on the other hand, is ambling calmly toward you. He casually raises a hand in greeting. "Ciao."

"No, I have not yet eaten, thank you." *Funny that he should ask*, you think. "Is Mrs. Brewster okay? Do you think the Kuritin —"

"Yes, yes. She'll be fine. Her problem is completely unrelated to the trial. Merely a coincidence. Not to worry, Parvesh." Crenshaw pats you lightly on the shoulder.

You draw the curtain to Room 4 aside slightly. Gloria is in the bed in the middle of what looks like a too-large room with a clutter of equipment pushed to the walls. Some of the devices you recognize, some you do not. Gloria looks colorless and drawn. She seems to have aged ten years since you saw her last. Walter and Wayne hover over her, silent and concerned. Wayne is giving her ice chips from a plastic cup.

You rap lightly on the wall. Both men look up. Wayne nods at you and Walter waves weakly. You enter the room. You figure Crenshaw is right behind you, but as you turn around, you see no one. You stick your head back out into the hallway just in time to see Crenshaw turn the corner and disappear. You walk to the foot of the bed. Gloria looks groggily at you. "Hi, Parvesh," she croaks.

"Hello, Gloria."

You are rewarded with a ghost of a smile. "You remembered."

"Yes, ma'am."

She lets it pass. "You're supposed ..." she says haltingly, "to bring a girl chocolates ... when you visit."

You can see why Walter was so concerned about her breathing. "I'll remember that next time."

Her eyes flutter shut. "I'm tired, Wayne."

"I know, Mama. You're gonna get some rest now." His voice cracks. Wayne is visibly holding back tears. You feel uncomfortable. You feel you should say something, but you are unsure what.

Dr. Vu enters the room, relieving the awkwardness. He tells Wayne, "She's going to be transferred to the ICU in a bit. We're taking her off the Kuritin until she gets her strength back. Probably for a week."

"What will that do to the cancer?" Wayne glances back at his sleeping mother.

"I've talked to Dr. Crenshaw about that. He agrees it's more important that she stabilize right now."

"Thank you, doctor." Wayne seems dazed, still looking at his mother. Walter, on the other hand, seems a mixture of annoyance and worry.

Soon, orderlies arrive and begin the business of relocating Gloria and her attached machinery and tubes. You tell Wayne and Walter that you will keep yourself up to date on Gloria's condition, then you get out of everyone's way. No one notices when you leave the room.

You drive home, sad and thoughtful. Another night of drive-thru Tex-Mex. The curry will have to wait.

TURN TO PAGE 423 TO CONTINUE.

# Parvesh: Unreported

YOU DID NOT SLEEP WELL LAST NIGHT and you blame Tex-Mex food. It was likely the matters of the previous day that kept you awake, but the food is a convenient scapegoat.

You settle with a sigh into your workspace at the lab. Your pile of books, notebooks and iPad smacking down on the table with more noise than usual. This causes Dr. Patel to look up from her monitor. "Had a rough night, my friend?" she asks.

You tell her about the phone calls and the hurried trip to the emergency room. "Uh-oh. That's called an adverse event." Patel regards you curiously, then shuffles through the papers in the inbox next to her computer. "Hmm. It does not look like Dr. Crenshaw has filed the adverse event report yet. Well, he has a week or so until —"

"He said it was unrelated to the drug trial."

"You mean … Dr. Crenshaw is *not* going to report it?"

"I don't know. It didn't seem like he would. He said it was a coincidence."

"A what?" Patel slaps her hand on the desk. "No, no, no, no, no, *no*. He has to report it." Patel rises. "Something like this *has* to be reported."

She glides past you toward Crenshaw's lab office. You indulge in another sigh and punch up the iPad, preparing for the day ahead of you. Your preparations are interrupted when Patel comes back into the room, doing her best to slam the door shut behind her. However, it squeezes shut silently on hydraulic hinges, which seems to annoy her even more. She is visibly shaken and angry as she resumes her seat at the computer.

You watch her, waiting. Finally, she speaks. "Do you know what he told me?" Patel is whispering. "He said it was *not* a serious adverse event and there was *no reason* to report Mrs. Brewster's reaction to the drug. He called it *unrelated!*" Patel brings her thumb and index finger together in front of your nose. "That man! I swear to you, that man was *this close* to telling me not to 'worry my cute, pretty little head' about it." Unconsciously, she adjusts her hijab. "I am beginning to see why the last research team left so suddenly." She swivels in her chair back to the computer screen. Windows begin to open and close, and she is typing furiously.

"What are you doing?"

"This *has* to be reported, no matter what Crenshaw says." She looks back to you. "The University has a whistle blower policy, yes? See here? It says, 'anonymity is guaranteed.' They will notify the IRB."

"Anonymity is *guaranteed?* Are you sure?" You are not so sure.

"Look, it even says, 'The compliance hotline has no caller ID function.'"

You are still not so sure.

"Parvesh, we *must* call the compliance hotline if Dr. Crenshaw will not report this. The more I know about the data on this project, the more I feel we are digging a very deep hole."

"We? I am only a first year dental student!"

"It does not matter. This is serious."

"I am not disagreeing, but he *is* the doctor —"

"I think there may be other problems."

"Like what?"

"Like, according to the data, almost 100 study patients have the same birthday."

You say nothing.

She glances at her computer. "There is only so much information I can access, but there is certainly enough to make me, or any reasonable person, suspicious."

"Perhaps we need some backup, as they say on television."

"You should not watch so much TV, Parvesh. Don't you know it will rot your brain?"

"Dr. Patel, you and I have to be careful. Especially you. You are on a visa. If we get ourselves in trouble, I could be kicked out of school and you could be sent home."

"So, I suppose you have a better idea?"

You do, or at least you think you do. "I met a woman at the beginning of the school year who may be able to help us. Well, at least, we talked briefly. I had wanted to … I mean … I never got the courage to ask her —"

"Stop your gibberish Parvesh. Who is she?"

"The university's Research Integrity Officer."

"I see. And how long do you think our anonymity will last with her?"

After several minutes of debate, you and Patel realize you are simply saying the same things over and over. Finally, there is a moment of pained silence. Then, Patel reaches for the phone. "This call is not going to make itself, Parvesh." She picks up the handset. "I am calling the IRB." You watch as Patel places the phone between her shoulder and head, then clicks her computer screen. The website for the university's Office of Compliance pops up. She scrolls down to a phone number and mailing address of the IRB office. The webpage lists two IRB chairs.

"Oh, no. It's him." Patel puts the handset down on the desk.

"What do you mean, Dr. Patel?"

"It's him, that jerk, Dr. Cochran. I completely forgot that he is one of the IRB chairs." Patel's expression has changed from anger to worry.

"Does it matter?" Fearless as she seems, you have no idea why this would bother Patel.

"It might. He hates me, and I hate him." Patel is frozen, looking at the screen. "I am no longer sure what we should do, Parvesh. What do *you* think?"

TURN TO PAGE 428 TO FIND ANOTHER WAY.
TURN TO PAGE 429 TO CALL THE RESEARCH INTEGRITY OFFICE.

# Parvesh: Call the IRB

PATEL CLICKS THE "PREVIOUS PAGE" BUTTON and her screen returns to the university's Office of Compliance. "It does not matter, Parvesh. We are anonymous, anyway." Patel taps the number into the phone. After a few rings, she speaks to what is apparently an answering machine. She gives a concise explanation of what you two have been discussing for twenty minutes, then hangs up. Your names are never mentioned.

"Well," she says. "There you are. I think we should be getting our data in a row, as it were," and she returns to her computer with newfound purpose.

If you had expected to find any relief in this process, you would have been disappointed. Instead, you are experiencing exactly what you thought you would: a sense of foreboding, of imminent doom. *What did we just do?*

TURN TO PAGE 432 TO CONTINUE.

# Parvesh: Find Another Way

YOU REALIZE PATEL'S PROFESSIONAL FUTURE in this country, not to mention possible deportation, cannot be taken lightly. You assume there must be a safer way. "Dr. Patel, Perhaps we should let sleeping dogs die."

"I believe you mean to say, 'lie,' Parvesh, and that is exactly what my fear is."

"Death of dogs?"

"No, death of people. Let me show you something." Patel's keyboard springs back to life. The next image on the screen is a page from Wikipedia. "Have you ever heard about Jesse Gelsinger?"

"No."

"That is the name of a patient who was enrolled in a gene therapy study some years ago."

"Yes?"

"He died in an adverse event, which was reported, but there were earlier adverse events that were unreported. Had they been reported, Mr. Gelsinger might be alive today."

You read the web page and experience a tingling sensation in the back of your head that is not pleasant. "Dr. Patel, we have to report this to the IRB."

"Yes, I believe we do."

TURN TO PAGE 427 TO CONTINUE.

# Parvesh: Call the RIO

YOU NOW REALIZE that running into Cheryl at the ER was classic karma in motion: she jogged your memory about Dr. Margeaux Frey.

As you pull up the university directory on your iPad, you tell Patel about the "cute redhead," as Cheryl put it, though you omit the rest of the conversation. "She has a degree in bioinformatics, so she will be familiar with data mining. I suppose, as the university's Research Integrity Officer, she will be getting involved in this anyway." You glance up to Patel, hoping for confirmation. "Yes?"

"I guess." Patel stares briefly into space, then back at you. "You say you know her?"

"Barely. Socially. Briefly." You are feeling flushed again, and that annoys you. "No. Well, a little."

Patel studies your face. "Then you, my friend, get the honor of calling her."

You touch the number into your iPhone and find yourself speaking with Frey's receptionist. You tell her that you need to see Dr. Frey in her RIO capacity, and your luck is either cursed or blessed, because she has a hole in her schedule almost immediately.

You grab your iPad and notebook and start for the door. "Good luck," Patel calls after you.

You stop and turn around. "You are not coming with me?"

"I sense three's a crowd."

"I am not brave enough to do this alone, Dr. Patel."

She smiles, then picks up her handbag and logs off the computer. "So, we are a team."

✵ ✵ ✵ ✵

*Oh my goodness, she is even more beautiful in street clothes.* That is the thought you try to squelch as Dr. Frey greets you in her office, that beautiful musical accent lilting in her welcome. You sit in the offered chair. Patel's nod to you communicates that you will start things off. You explain thoroughly and quickly the events concerning Mrs. Brewster. You stammer at first, but your delivery becomes easier as you see Frey is genuinely listening to you, and making notes.

"So … you two think it likely that the drug caused her seizure yesterday and that it was due to Mrs. Brewster's age?"

"From my literature review and reading of the IRB protocol, yes. I … I mean we … we noticed Mrs. Brewster's health had been deteriorating since the treatments began. She had a clean bill of health before, except for the lesion."

"Parvesh, I can't help noticing that you are a first-year dental student. Are you providing me a medical analysis here?"

"No?"

"You know that Dr. Crenshaw has ten days to file a serious event report. He still has time to do that."

Patel speaks up. "He will not."

"How do you know that, Dr. Patel?"

"Because he told me."

"Please go on."

Patel doesn't stop talking for five minutes. After that, there are a few more minutes of exchanging contact information and particulars. Then Frey says, "Well, Parvesh and — it's Faiza, isn't it? — you were right to bring this to me."

Mainly you just hope she will say your name again. You leave her office and return to the lab, feeling a bit lighter. You feel that you have made a powerful ally, and the path ahead is no longer quite so dark.

TURN TO PAGE 432 TO CONTINUE.

# Parvesh: Investigation

IT HAS BEEN ALMOST TWO WEEKS since you and Dr. Patel made your "Fateful Decision" — calling the university's compliance hotline to bring your doubts about the research trial into the light. It has seemed to make little difference in the interim. The two of you still report to the lab every day, you gathering data from study participants and Dr. Patel entering data into spreadsheets, then grudgingly forwarding a copy to Dr. Crenshaw to "fix holes."

Today, however, is different. It's early in the morning when you hear the door into the lab open and close behind you. *Who could that be?* Dr. Crenshaw usually doesn't come in until after lunch on Wednesdays. He spends most mornings in his other office off the main lobby, the one he affectionately calls, "The Palace." Patel looks up from her monitor and stiffens. You slowly turn to see a man standing in the doorway, a man who could easily be mistaken for a Hollywood actor playing the role of "The Good Doctor." Everything about him seems straight out of Central Casting: a ruggedly handsome Robert Redford type sporting a mane of slightly graying hair. You notice his crisply pressed white coat has cloth knot buttons, not plastic ones — a subtle sign of power. You also notice The Good Doctor is not happy.

"I might have known," the man growls.

"Dr. Cochran," Patel states flatly.

"Faiza," Cochran rejoins, striding further into the lab. "The IRB received an interesting phone call this morning. It was from a lawyer, which isn't too unusual for the IRB, but I'm informed that this call was made in reference to the Kuritin trial being run from this lab. I am also aware that two weeks ago, an anonymous call was made to the university's whistle blower hotline about the same drug trial. Now … there are only two people I know of working on this trial, other than Dr. Crenshaw: his research coordinator and his research assistant. Perhaps you can tell me where I might find those two?"

Patel speaks up, "We are concerned about gross —"

"I get that," Cochran interrupts. "I totally get that." Ignoring you, he walks menacingly toward Dr. Patel and sits on an empty stool next to her. His nose is only 6" away from Dr. Patel's. "Look Faiza, let me pay you the return favor of being blunt." Cochran folds his arms. "You know epidemiology and you know research, of that I have no doubt. However, you have *no concept* of politics, and at this level of academic life, research is as much about politics as it is data."

"What are you saying?"

"What I am saying is that Dr. Crenshaw has been the linchpin of the university's Crenshaw Center since I started here twenty-five years ago. Did you notice his grandfather's name above the front door when you walked into the building this morning? Did you happen to see his eight-foot tall portrait in the lobby? Or the word, 'Crenshaw' on the hundred or so donor plaques by the fountain?" Cochran stands up. "Open your eyes, Faiza. You and Parvesh over there are sitting in a 150 million-dollar facility that supports a multimillion-dollar research program and enormous clinical enterprise. The amount of new donor support the Crenshaw Center receives every year is also in the millions. You two monkey with the power and reputation of a name and place like this, and it's going … it's going to look like …" Cochran stops mid-sentence, thinking.

Patel says simply, "Say it."

"It's going to look like a couple of clueless foreigners trying to take down a large and well-respected American institution. This cannot end well. For you two, I mean."

There's the sound of another door opening, this time from Crenshaw's lab office. "How about me, David?" comes a sweet voice in the doorway. "I'm Canadian. Am I one of your evil foreigners too?"

Margeaux Frey, bless her, is standing in the doorway, a number of folders clutched to her chest. *Where did she come from? Has she been in his office all morning?* You seem to imagine a halo around the red curls of her hair, but it's likely due to the fact that you have been holding your breath for the last several minutes and are close to passing out.

"Margeaux," Cochran acknowledges. "So they've pulled you into this, too?"

"I would not say, 'pulled,' David. You are aware that Dr. Crenshaw has not reported a serious adverse event in his study within the time limit?"

"That was the subject of the phone call this morning. I haven't had any time to —"

"No need," Frey interrupts, handing Cochran one of her folders. "I have had two weeks to do a little research on Dr. Crenshaw's research. Did you know that this was not the first time a complaint has been registered about one of his trials? And that one of the other complaints came from his previous study coordinator?"

Cochran sits down on the stool. He places the folders on Patel's workstation. One by one, he picks up a folder and starts reading. "These are all before I was instated on the IRB."

"And before I was appointed RIO."

Cochran looks up from the paperwork. "None of these ever got to the inquiry stage?"

Frey shakes her head. "No. As you say, he is an important man here. Perhaps important enough for things like these to be dismissed without investigation."

Cochran hands the pile of folders back to Frey. He sits on his lab stool and stares toward Crenshaw's office. You are reminded of a ship's sail suddenly without wind to support it. He looks up to Frey again. "You're serious about this?"

She considers him. "The adverse event is a fact."

Cochran turns to you and Patel. "Is it?"

You finally find your voice. "Yes … Yes, sir, it is. I was at the Emergency Ward."

Cochran silently nods, then swivels his head to Frey. "You're taking this into an inquiry?"

"I am about to gather evidence, then we will see. We spent this morning running the study's lesion photos on Dr. Crenshaw's computer through photo-authentication software, and I did not like the results. My IT person is in Dr. Crenshaw's office now with his computer. Campus Security is in the hall."

You look past Frey into Crenshaw's office and notice the back of a person hunched over Crenshaw's desk near his computer.

"Incidentally, Faiza," Frey says. "I need to ask you to step away from your computer."

With the slightest of smiles, Patel raises her hands and pushes her chair away from her workstation.

Frey turns back to Cochran. "The rest of my Hit Squad is on the way to pick the remaining computers in the lab."

"I wish you wouldn't call them that," grumbles Cochran.

"After we've gone through the data, I can tell you if an inquiry is necessary. I also believe that you and I, David, should meet with the adverse event patient and investigate that issue further."

"She will probably bring the attorney I spoke with this morning."

"Fine by me."

Cochran rises from his seat. He considers each one of you in turn. "Just remember what I said. The Crenshaws and their clinic have brought hundreds of millions of dollars into this university, not to mention a ton of prestige. There is no way this will be quiet or genteel. You are talking about unleashing a hurricane."

"David," Frey says softly, "there is a possibility that he has been gambling with patient lives. Should we just ignore that? *Can* we ignore that?"

Cochran is silent for a moment. Then, "No. No, of course not."

He walks to the door. He turns again to face the three of you. "I feel like I should say something like, 'Good luck,' but I don't know who to say it to."

TURN TO PAGE 439 TO CONTINUE.

# Pool Party

LONG SHADOWS POINT toward an orange sun slipping behind a quilted landscape. The golden hour light must be a sign. It is late May in Austin. The year's unusually cloudy spring is grudgingly giving way to summer. The change of season announces hotter days and warmer nights — a perfect opportunity for another pool party at the home of Dr. Enrique Hernandez.

One again, his backyard is full of people. Once again, the sounds of revelers echo off suburban hills. However, this time, there is a difference even casual observers would notice. There are no bathers in the pool. Instead, the patio is an ocean of humanity — and a turbulent sea, it is. A great number of guests are moving about in white coats, giving the impression of swirling rapids encircling the yard.

There is a greater disparity of ages here than at the pool party nine months ago. The very young are running about, some laughing, a few whining, and one is screaming. The very old look on tolerantly. The middle aged speak to the white coats in faces beaming with pride.

A clanging of glass quiets the crowd and ceases the band's aimless tuning. The host steps up to a microphone. He looks over the assembly and throws his hands in front of his face. "Whoa! I should'a brought my sunglasses! Look at all that white out there! I am snow blind."

General laughter accompanies the crowd coalescing in front of the band. Hernandez smiles and relaxes into a casual stance, his hands in his pockets. "I want to congratulate each and every one of you on receiving your white coat today. This is an important day for you. On this day, you passed from the preclinical to the clinical. You've got some basic science down. Soon you will start learning how to use that science."

A smattering of applause turns in to rancorous clapping.

"I know some people don't like the idea of white coat ceremonies. They say it's like a priest being ordained into a secret cult, that it breeds elitism, a sense of entitlement." He shakes his head. "Well, that may happen to some of you. I hope it doesn't. If it does, I look forward to addressing you as Dean."

More laughter. Hernandez makes a show of scanning the crowd. "Um, if the Dean is here this evening, please tell him I'm just kidding. *Please?*"

A male voice calls from the house, "No!"

"Uh-oh," laughs Hernandez. "He *did* make it!"

When the clapping dies down, Hernandez continues. "I also want to congratulate those of you *not* wearing white coats. The family, the friends. The people who have loved and supported these students through some rough times, physically, emotionally, and financially. Today is your triumph, too. This is *your* celebration as much as anyone's. These guys and gals are lucky to have you."

A groundswell of applause, but this time only from the white coats. Hernandez looks gratified.

"White coats, many of you took the Hippocratic Oath this morning. Others recited the Nightingale Pledge. There was an oath for each health professional school in the university.

Whichever one you took, consider this: you made a solemn promise today — but your oath is much more than that. Your oath is also a guidebook on how to conduct your professional life. Always remember that, and always remember this day. Today, you took the next step into a brave new world."

He looks over the crowd, drinking in the expectant silence.

"Now let's party."

There is a great roar as Hernandez yields the microphone back to the band. A man with a guitar slung over his neck lowers the stand. "Good evening everyone. I'm Larry Wayne and we're Larry Wayne and the Wranglers. I hope y'all like country music mixed up with a little rock and roll."

Cheering.

"I was told to tell you that the barbecue line is now open. Dinner is served."

Hoots and hollers.

Larry is working the crowd. "I can see you're in a festive mood. Dr. Hernandez said there are parents here tonight. For y'all, we'd like to begin our first set with some oldies but goodies. This one is from my favorite 70's bands. Anyone here old enough to remember The Kinks?"

✳ ✳ ✳ ✳

"Hey, they were also one of my favorite groups when I was your age." Wayne Brewster is looking at his son, Walter, who is smiling and modeling his new coat like a runway model.

"The who?" asks Walter.

"No, The Kinks."

"What?"

"Never mind." Wayne smiles at his son. "I gotta admit, Walter." Wayne brushes imaginary lint off his son's shoulders. "There were times when I thought I'd never see you in one of these."

"Yeah. Truth be told, me either. I still feel like I'm a kid play-acting in this thing."

"Sorry," says Wayne with a bit of rue in his voice, "one of the things you don't get to claim anymore is being a kid."

Walter stiffens up and straightens his coat. "I don't guess I do."

"Fact is, there's been quite a change in you the last few months. Your mother noticed it right off."

Walter scans the crowd, his gaze finally settling on his mother way over near the house. Sheila is having what seems to be a serious talk with the nursing student who took care of him after the car wreck last August. "Many things got clearer to me, Dad … I guess. Like, if I was going to be a doctor … if people like Nani needed people like me …" He looks back to his father, "Well, I figured it was time to get serious about it." Walter and Wayne's eyes meet. "It's like I'm operating at a whole new level now, Dad, and I like it."

"We can see that you do."

Walter looks back toward his mother. "I'm glad you two are giving it another go."

Wayne shrugs and glances in Sheila's direction. "We've got a lot invested in this marriage. Our parents didn't give up when things got rough. So, when your mother came to me and apologized —"

"You're kidding."

"No."

"I've never seen her apologize for anything."

"I have to tell you, Walter, she wasn't always that way. Your mother apologized … we talked … we cried … all of a sudden we were college students getting caught in the rain at that stupid Kinks concert again. Something magical happened, back then and now. I had my old Sheila back. I felt more like myself than I had in ages."

There is a brief silence as Wayne studies his wife and takes in the music, his mind obviously elsewhere. "Listen, they're playing the song your mom and I met over, *Alcohol*."

"Um, you want me to get you a beer or something, Dad? Some wine?"

Wayne shakes his head. "Nope. I haven't touched the demon alcohol since your mother left." Wayne is singing the chorus line. His mouth turns into a devilish smile. "You know, that song may be the reason you're here in this world. Ah well, that's another story."

Walter grins.

Wayne turns a little serious. "No. Having to look after the girls, then your grandma getting sick, well, having a drink just didn't seem …" His voice trails off, and he shrugs again.

"You miss it?"

"The old times? Being young? You bet. A drink? Sometimes. Not often. It was starting to feel like booze was running my life, and that was pissing me off." Wayne motions to the Dr. Pepper in his son's hand. "You?"

Walter shakes his head. "Not since the wreck. I guess that's another change I racked up this year. It's sure made studying easier."

The men grin at each other, not sure what to do next. Wayne finally says, "I should go check on your mom."

"I'll track down Nani."

Walter breaks into a wide smile. "We both know she's probably teaching some students line dancing right now."

Wayne, chuckles. "Yup. Or more likely bragging about her grandson, 'the medical student.'" Both men laugh. "Walter, your Nani sure is proud of you. She's fit to pop."

Walter looks down at the ground. "Yeah, well …"

The silence returns. Both men almost say something, but don't. Wayne's smile becomes bittersweet, a smile with the weight of memories. "My son the doctor," he says softly.

"Not yet, Pop, but working on it."

Wayne surprises them both by lurching forward and embracing his son. Slowly, Walter realizes that this isn't going to be a quick but embarrassing squeeze. Walter's arms come up and return the hug. Wayne speaks directly into Walter's ear, "Your Nani ain't the only one proud of you, my boy." The dad steps back and gently taps the son on the arm.

Walter's eyes close. This is something he's waited a long time to hear. Now it's Walter's turn to hug. "Thank you, Daddy," his voice cracking like a child's. "Thank you."

The men slowly separate. Wayne makes an aimless motion toward the party. "Need to check on your mother," he says.

"Yeah," answers Walter. "I'll see you."

Two men go separate ways, both making surprisingly identical gestures to brush at their eyes. Must be a bit of pollen in the air, or perhaps errant bits of hill country dust. Yes, surely that was it.

## * * * *

Julie Spates would rather be standing in the long line in front of the barbecue table. In fact, she'd rather be anywhere but here. Sheila Brewster saw her immediately after arriving at the Hernandez "hacienda" and pulled her to the side of the yard for a "private conversation." It began with, "You and I have some unfinished business, Julie." It's been the riot act since.

"Now, a little Paul Simon, by request." The Wranglers' drummer kicks off a rhythmic, syncopated beat using only the rim of his snare drum and his base drum's foot pedal. Hands start clapping. Someone nearby says, "I haven't heard *Fifty Ways to Leave Your Lover* in years."

Sheila pulls Julie by the arm closer to the house, away from the music.

"Mrs. Brewster, I have *no idea* why you think I betrayed your confidence. I swear to you, I said nothing!"

"So you told me. But I can add, Julie. I can put two and two together and get four. Thanks to you, my marriage almost broke up. Despite that, I managed to patch things up with Wayne. But, I will *never* forgive you for what you did to me."

Julie is exasperated. "Mrs. Brewster ... I ... I ..."

Carla Hernandez knows trouble when she sees it. Too many years in the courtroom have made her an expert at reading faces. Right now, the face on her friend Sheila outside her front door spells *disaster*. Time for an intervention.

"Are you two having fun?"

"Carla, you're just the person I was looking for. I should have come to you months ago. I should not have kept silent." Sheila's grip on Julie's arm grows tighter. She pulls Julie to the door.

"Something wrong?"

"Julie — that's what's wrong. She told my husband about a little fling I had and it almost wrecked my marriage."

"It's not true, Mrs. Hernandez. I never —"

Carla pries Sheila's hand from Julie's arm. Julie starts rubbing an area she's sure will be sore by tomorrow.

"Sheila, are you talking about Armand, your boss?"

"Oh for crying out loud! You know *too*?" Sheila's face turns red. She looks over the patio of people. "Is there anyone at this party who *doesn't* know?"

"Calm down Sheila. Calm down." Carla looks at Julie and winces. "I'm afraid there may … I think this is a case of mistaken identity, Sheila." A tear forms in Carla's eye. "Julie, please leave us. Sheila and I need to talk."

\* \* \* \*

Parvesh and Patel stand at the low wall at the rear of Hernandez' back yard, taking in the hillside view.

"The white coat looks good on you," Patel says, sipping from a cup of juice. "Very professional."

"Thank you, and thank you for coming, Dr. Patel. It means a lot to me."

"It's too bad your family couldn't be here for this."

"I wanted them to come, but my mother was not up to traveling from New York … but they will definitely be here when I graduate."

"Then, I am honored to be serving in their place for you today."

Parvesh smiles. "Yes, you are my *research* family." He looks around Hernandez' patio, at the other white coats surrounded by family and friends. "I must tell you, after spending so much time splitting my days between the research study and classes, it is a real luxury to have mornings free. Most of my professors commented on how much I improved afterward. I learned valuable research skills from you, Dr. Patel. I am very thankful."

"Oh, don't thank me. You are a quick learner."

"So, what have you been doing since they shut the trial down?"

"I have been attached to another project." Patel steps closer to Parvesh and raises her hand to cover her mouth. "To tell you the truth, I was afraid there might be some fallout from the Kuritin disaster, but I am told I came highly recommended."

"That is wonderful! I hope you are happier there."

"Much. I was lucky. I am still trying to figure out who put the good word in for me."

"Hello there!" comes a jocular voice from behind Parvesh. He turns to find Dr. Cochran with a bottle of white wine in one hand and a full glass in the other. His tie hangs loosely from his collar. He looks quite relaxed, and very tanned. "Far away from the maddening crowd, eh?" Cochran gestures with his bottle toward Parvesh's cup. "May I?"

"Oh yes, thank you."

He pours a generous amount of wine into Parvesh's cup, then turns to Patel. "Dr. Patel, I won't insult you by offering."

"I should think not," she replies cautiously. "I had not expected to see you here."

"I never miss one of Enrique's parties." Cochran takes a sip from his glass and looks at the hills. "It's really beautiful out here. I think he's got over 3 acres."

"Yes, very."

Cochran stares at the sunset. "I have to say, Faiza … the Crenshaw affair ended about as well as it could."

"So he's gone?" asks Parvesh.

Cochran turns back. "We offered him a package. Crenshaw stepped down last week. After much hemming and hawing, Dr. Nguyen was appointed interim director. The university would prefer a big name to take over, but she'll do fine. I hope they make it permanent, but my guess is we'll see a national search before long."

"What will become of Dr. Crenshaw?" asks Patel.

Cochran shrugs. "He'll land on his feet. By resigning before there was a formal investigation, his reputation is still intact. The sad thing is, he was getting good results with the drug."

"But," says Patel, "he invalidated it with a bad protocol."

"It was not a bad protocol, Faiza. Crenshaw had an excellent protocol. He just ignored it." Cochran looks toward the pool. "I think I saw the Brewster woman here. I need to talk to her. There's another Kuritin trial being put together at the Medical Center in Houston, but this one will be geared to older patients. I can pull some strings and get her in, if she likes."

Parvesh nods. "Her lesion was shrinking, that part of the data was true. I know her a little better than you, Dr. Cochran. Perhaps I should break the subject to her?"

Cochran smiles. "I think the word is 'broach,' and yes, that's a good idea."

Parvesh sets out to find Mrs. Brewster.

Patel considers Cochran for a moment. "Speaking of pulling strings, Dr. Cochran …" Patel hesitates, hoping Cochran will finish her sentence. He doesn't. "Dr. Cochran, did you pull some strings for *me*?"

"You did good work on the trial, Faiza. Well, such as you were allowed to do. I see no reason why you shouldn't go on doing good work."

"I have to admit, Dr. Cochran, I did not like you. Chances are good I still do not like you very much. But it cannot be denied that you can do the right thing. Thank you for your recommendation."

Cochran rotates hillside again and raises his glass to the setting sun. "Dr. Patel, here's to you."

"What?"

He turns and gently places his hand on her shoulder. "I want you to know, Faiza, that I admire your chutzpah. You did a brave thing. I don't think I would have had the … guts … had I been in your place. Here's to you and to Parvesh."

Patel is taken aback. "I … I don't know what to say."

Cochran gets a distant look in his eye, obviously speaking from memory. "The world and all things in it are valuable, but the most valuable thing in the world is a virtuous woman."

Patel is obviously shocked. "You *have* read the Qur'an!"

Cochran nods. "More than you give me credit for. I hope that you will forgive me if I stick with my own prophet."

"Of course," Patel smiles. "Jesus was a perfectly good prophet."

"Moses, actually." Cochran raises his glass in another toast. "Keep up the good work, Dr. Patel. L'chaim!"

"Excuse me?"

"To life!"

Patel raises her cup of juice. "L'chaim to you too, Dr. Cochran."

**\* \* \* \***

"Thank you, thank you." Larry Wayne is changing guitars among a smattering of applause, feedback squeals and input pops. "Okay, time for a little country music."

"Yeehaw!" comes a single voice from the crowd, followed by a round of chuckles.

As if on cue, Larry Wayne becomes Willie Nelson singing, "Mamas, Don't Let Your Babies Grow Up to be Cowboys."

Another lone shout from the middle of the crowd. "SING IT, WILLIE!" Then, a long, two-finger whistle. Gloria Brewster drops her hand to her side.

"Nani!" I had no idea you could do that."

"Shoot, Buddy. You ain't heard nothing."

Walter starts to respond but is quickly silenced.

"*Shhhh!* Here comes my favorite part."

A few seconds later, Walter asks, "You mean the part about making them be doctors?"

"Yep."

"What about lawyers and such?" Rachel Parker appears from behind a group of white coats.

"Rachel, I didn't know you were coming," says Gloria. "It's good to see you. I must say, you look darling in that blue dress."

"Oh, thank you. I wanted to wear jeans, but Walter insisted I wear this."

"*Did* he?" Gloria beams. Walter shuffles, then takes Rachel's hand. "Dance?"

"Sure."

Walter and Rachel disappear just as Wayne arrives arm-and-arm with Sheila. "Got us all Dr. Peppers, Mama."

Gloria takes her cup and points it toward the only couple dancing in front of the band. "Sweet, huh?"

"Yeah. You done well, Mama." Wayne and Sheila look at their son, then at each other.

Parvesh joins the group. "Hello everyone."

"Parvesh! How *are* you sweet boy?" Gloria throws her arms around him.

Almost spilling his drink on his white coat, "I am great. And you … you look so pretty today."

"I *feel* pretty. I also feel a lot better thanks to you, and a few others." Gloria winks at Wayne.

Wayne introduces his wife to Parvesh. They shake hands, somewhat formally.

"Parvesh here is my main man, Sheila." Gloria pinches Parvesh's cheek. "I'm going to adopt him."

Parvesh breaks into a wide grin. "Gloria, I know you are probably sick and tired of studies, but I want you to know that another Kuritin trial is starting. It is designed for older patients."

"Never call Gloria 'old,' Parvesh," Sheila jokes.

"Oh, I didn't mean it that way. I mean, it is for mature patients. Anyway, Dr. Cochran says that he can get you in if you like."

"My friend," Wayne says, "I believe we've all had enough of Kuritin."

"Thank you Parvesh." Gloria raises her index finger for emphasis. "*I'll* think about it. Right now, though, I'm living one day at a time, and trying to make the best of every one them."

"Yes, I can see that."

"Hey! Let's forget about all that for now." Gloria motions to the line, which is finally getting shorter, at the barbecue table. "I'm hungry. Who wants to fix me a plate?"

"We'll get it, Mama." Wayne and Sheila start for the house.

"Extra chili con queso! And bring me a couple of jalapeños too."

Parvesh and Gloria stand alone, looking at each other, smiling. "You look mighty handsome in your white coat, Parvesh." Gloria scans the crowd. "So tell me, Parvesh, have you found your soulmate yet?"

* * * *

"You guys are great." Larry adjusts his mic stand as the applause and whistles dies down. "Me and the Wranglers have to take a short break while I fix a guitar string. Be back in a jiff."

John and Cheryl sit on plastic chairs by the poolside. John is in scrubs, having just finished a stint in the ER. Cheryl, in stark contrast, is in a bright red dress. Both look rather out of place amongst the white coats.

"Cool dress, Cheryl."

"Cool scrubs, John. Were they expensive?"

"Free. Yours?"

"Ralph Lauren. I splurged."

"Shame on you."

"How often do I get to wear a dress anymore? Not much, I can tell you." Cheryl sighs. "Seems that I don't go to as many parties as I used to either."

"*Ha!* Who does?"

"So why aren't you wearing *your* white coat, John. It would have completed your ensemble."

"Meh. It's in the car. Getting tired of looking at it. So where is yours?"

"At home. Would have messed with the lines of my dress." Cheryl steps back and does a 360-degree turn.

"I already *said*, it's a cool dress."

"Yeah, yeah." She surveys the students surrounding them. "Look at them John. They're so ... *young*."

"Haven't we had this conversation before?"

"Were we ever that young, John? And that proud? And that *sure* of ourselves?"

John strikes a dramatic pondering pose. "I believe we were. I believe some of us are *still* that young."

Cheryl tosses him a long-suffering look. "I have *got* to stop lobbing you softballs."

"You pitched it, so I hit it." John holds out his hand to Cheryl. "Look, we're both still young, and so is the evening. You want to leave these starry-eyed kids and get some coffee?"

A guitar chord wafts over the crowd as the band returns.

Cheryl takes a step back, looks at John, at his outstretched hand. "Why, John Guerra. Are you asking me out on a *date*?"

"Oh, no. I'm too cheap for that. It's *coffee*."

Another chord, followed by Larry launching into a great impersonation of Ray Charles singing *Georgia On My Mind*.

Returning the smile, Cheryl takes the hand and pulls John out of his chair. "Nah. Just wanna dance. Come on, they're playing our song."

Cheryl and John make their way through the crowd to the small clearing near the band. The music plays as a lone pair, one in white and the other in blue, circle to the beat. A man in green and a woman in red join them. Soon, a few more white coats drift in, and then a few more, and a few more … It isn't long before the growing crowd blurs into a swirling sea of synchronously moving white waves.

TURN TO PAGE 457 TO READ RESEARCH ETHICS.

# 3: Research Ethics

# 3.1: The Scientific Method & Research Integrity

THE SCIENTIFIC METHOD is a systematic process of observation and experimentation that results in measurable evidence. It's the foundation for obtaining valid results in clinical research, and the basis for many ethical issues in scientific research.

Science is concerned with discovering new knowledge, which means not only arriving at the truth but getting there using the right methods. Hypotheses must be formulated and tested. Conclusions must be supported by results. High value is given to skepticism in science. It is better to honestly identify questions left unanswered than to assert greater certainty than evidence supports.

This view of scientific methodology was originally developed in the natural sciences. Today, it applies equally to the quantitative biomedical and social/behavioral sciences, even though the social/behavioral sciences are more likely to test stochastic or statistical correlations rather than causal relationships. Before beginning any research project you must conduct a thorough literature search. This avoids repeating work already done, helps develop the most sophisticated hypothesis to test, and maximizes the chance of discovering new knowledge. A literature search is also important so that those who have worked in the area receive the recognition they are due. To fail to cite important work that helped develop a research project or protocol, sometimes called "citation amnesia," is unethical because it denies others fair recognition. For any study, results must be repeatable by others. Methods and data, not just results, must be systematized and made public. This may require, in biomedical research for example, that you be willing to share essential reagents or antibodies with other labs (with reasonable protections for intellectual property claims).

When learning to become a scientist, you learn these skills and behaviors by being part of a laboratory team directed by an already established scientist. Mentoring is part of this socialization process. Established scientists at universities and centers around the world have a responsibility to teach and promote development of the next generation of scientists. Senior faculty not only help students develop their research technique, but are role models for success in the field. They are ethically bound not to abuse their power by giving students menial jobs unrelated to their education (e.g. getting them coffee, or babysitting their children), or keeping them in school longer than necessary for their education to get more work out of them. The most extreme form of abuse is sexual harassment. As noted in the instructional materials on Professionalism, any unwanted sexual advance, comment, or innuendo could be a form of sexual harassment. Sexual harassment is illegal as well as unethical and can be reported to a department chair, dean, or compliance officer.

Conflicts of interest were addressed as an ethical issue in the section on Clinical Ethics. COI is an equally important issue in Research Ethics. In research ethics the focus is on biases that might lead to inaccurate publications. While financial conflicts are commonly recognized, religious and political beliefs can also bias your research. The danger is that various political or personal interests will take precedence over objective reporting and interpretation of data.

Because of the nature of the topics they study, social scientists must be especially careful about these dangers. Self deception, the denial of the influence of personal beliefs or values, is more subtle and perhaps more dangerous than conscious forms of bias. Examples of religiously or politically charged issues that might challenge objectivity or encourage self-deception are: gun control, abortion, global warming, environmental protection, homosexuality, racial differences, and illegal drug use. A general rule for recognizing one's biases might be: If you know in advance what you hope to prove, you are at risk for biased research. Getting results you did not expect probably indicates unbiased research.

# 3.2: Authorship

THERE WAS A TIME when most scientific articles had a single author who did all the research. But modern scientific research is more complex, requiring a number of collaborators and authors contributing to research manuscripts. Peer reviewed scientific journal articles now typically have three to eight authors.

As a consequence, the problem of who should be listed as an author has become an important topic in research ethics. Over the last 30 years, a consensus about criteria for being included as an author has been developed by the International Committee of Medical Journal Editors.[62] While practice varies by discipline, the general idea is that each author should make a substantial (quantity) and significant (quality) contribution to the research; that this contribution should be conceptual (in study design or data analysis); that each author should have drafted parts of the grant or manuscript (and preferably both); should have read and approved the final version of the complete article, and be able to respond to questions about the work. Many journals require authors attest to these facts before their article can be accepted.

Not all collaborators should be listed as authors. Some deserve credit, but as an acknowledgement. Contributions such as provision of materials used in the lab, performance of assays, use of facilities, or routine patient care do not alone justify authorship. Participating solely in acquisition of funding or in data collection also would not suffice. Authorship should not be a gift for in-kind contributions or to help promote a career.

Being added as an author of a research paper you were not closely involved with is unethical, unprofessional and potentially embarrassing. It is possible that the authors committed FFP (fabrication, falsification, plagiarism) or some other type of misconduct.

The primary author (usually the first listed on the manuscript) takes responsibility at the outset to insure all members of the research team understand and agree to these authorship principles and see that all authors approve the final draft. Contributing authors must accept the responsibility of avoiding unnecessary duplication of journal publication of similar material.

There is an obligation to make sure everyone who made a contribution is properly identified and credited. Leaving out an author who deserved credit is unethical, no matter the reason. Even if a student, post-doc, or faculty member has left the lab or the university, they still deserve credit for their work. Readers deserve to know who really wrote a paper. Many articles have been published using ghost authors paid by a pharmaceutical or medical device company sponsoring research, which may produce biased data or interpretation. Readers cannot fairly evaluate an article if authorship is not accurately disclosed.

Sometimes a sponsored research contract includes a right to prevent publication in case results are not supportive. This undermines the objectivity of science as well.

It is a financial conflict of interest for a scientist to allow private funding sources to obtain power to withhold negative results, or to hire a ghost writer to make results look better (data polishing). It is also important not to allow ghost writers to write the abstract, since many health professionals have time to read only the abstract. Many people working in poorly funded areas of research, such as public health officials and non-government organizations, often cannot afford expensive journal subscriptions and make do with publicly available abstracts. Journal editors now force authors to retract articles when conflicts of interest are exposed after publication. It is common to require a full disclosure before the article is reviewed, with any conflicts of interest disclosed in a statement alongside the article.

# 3.3: Misconduct, FFP, & Data Management

WHAT IS KNOWN AS "RESEARCH MISCONDUCT" today was once called "FFP," for Fabrication, Falsification, and Plagiarism. Fabrication is making up data or results and recording or reporting them as if they happened. Falsification is manipulating research materials, equipment, or processes, or changing or omitting data or results such that the research is not accurately represented in the record. It differs from fabrication in being more of a deliberate distortion of the facts rather than pure creation *ex nihilo* (which means, created out of nothing). Plagiarism is the appropriation of another person's ideas, processes, results, or words without giving appropriate credit. All count as cheating.

Plagiarism is a common research ethics problem. This topic is included in the Professionalism section ("Academic Integrity/ Academic Misconduct"). Two key points to remember: self-plagiarism is never permissible, nor is cutting and pasting or paraphrasing other sources (published or unpublished, famous or obscure) without attribution.

What began as FFP and was then called research misconduct is also more positively known as Responsible Conduct of Research (RCR). Whatever the label, the field of research ethics began with and gets its energy from misconduct. There are a dozen cases from the 1980s that are used as examples of FFP, including one where a perpetrator went to jail.[63]

Besides the Wakefield study on vaccines and autism described in chapter 2.16, another fairly recent case exemplifies key issues. In 2000, Werner Bezwoda, PhD was accused of falsifying research conducted in the 1990s intending to show high-dose chemotherapy and bone marrow transplantation prolonged the life of women with advanced breast cancer. According to press accounts, tens of thousands of women had undergone the controversial procedure, with 10% to 20% dying from the treatment.[64,65]

These are but two recent cases revealing the potential for research misconduct to violate the principle of Non-maleficence and cause harm to patients. For scientific and clinical reasons, it is essential to follow the rules of Responsible Conduct of Research. This includes keeping good records of all research and keeping it in a format that cannot easily be changed. Sewn-bound lab notebooks with data recorded in ink on consecutive pages is traditional, although electronic records can also be used. Loose leaf notebooks is highly discouraged.[66] All data should be accurately recorded when measurement takes place (or the same day). Omitting a result is falsifying data. A note should be written whenever a data point is dropped that includes a good scientific rationale.

Similar rules apply to photos and images. With digital photography and manipulation software like Adobe Photoshop it is possible to distort images to tweak data. This can apply equally to biological and social science research. There is a fine line between acceptable enhancements and scientific misconduct. Do not succumb to your own desire or pressure from others on your team (not even the principal investigator) to falsify data or alter images to get a specific result.

If you feel like you're being pressured, consider confronting the person or talking to someone you trust about whether you are reading the pressure correctly. If you conclude something unethical is being encouraged, report it to the principal investigator (PI), department chair, dean, or the Office of Research Integrity (ORI). The ORI of the Department of Health and Human Services (DHHS) has developed an interactive educational movie called "The Lab" for ethics education regarding research misconduct.[67] The film is a choose-your-own adventure, much like *The Brewsters*. It's both educational and entertaining. Anyone working in a lab should watch it. We also encourage students to review other sections of the DHHS ORI website for further educational resources, information, and opportunities.

# 3.4: Nuremberg & Informed Consent

DURING WORLD WAR II, German doctors performed many heinous, cruel and often lethal experiments on unwilling civilians in institutions like nursing homes, mental hospitals, concentrations camps and in occupied areas.[68] After the war, twenty Nazi doctors were put on trial for "crimes against humanity" in the International Military Tribunal case: *U.S. v Karl Brandt et al.* After the trial, the Court wrote, "The Nuremberg Code" (1947), which defined ethically permissible research with human subjects. Of the code's ten statements, the first and often quoted sentence is, "The voluntary consent of the human subject is absolutely essential."

The importance of informed consent had been recognized by many researchers as far back as 1900. But racism, anti-Semitism, eugenics (controlled breeding to "improve" the human species), and pressures of war blinded German researchers to human rights and human suffering.

The full text of the Nuremberg Code's first statement is:

> The voluntary consent of the human
> subject is absolutely essential. This means
> that the person involved should
> have legal capacity to give consent; should
> be so situated as to be able to exercise free
> power of choice, without the intervention
> of any element of force, fraud, deceit,
> duress, over-reaching, or other ulterior

form of constraint or coercion; and should have sufficient knowledge and comprehension of the elements of the subject matter involved as to enable him/her to make an understanding and enlightened decision. This latter element requires that before the acceptance of an affirmative decision by the experimental subject there should be made known to him the nature, duration, and purpose of the experiment; the method and means by which it is to be conducted; all inconveniences and hazards reasonable to be expected; and the effects upon his health or person which may possibly come from his participation in the experiment.

The duty and responsibility for ascertaining the quality of the consent rests upon each individual who initiates, directs or engages in the experiment. It is a personal duty and responsibility which may not be delegated to another with impunity." [69]

Twenty years after World War II, Henry Beecher's article "Ethics and Clinical Research" (1966) in the U.S. and a similar article entitled, "Human Guinea Pigs" by Maurice Pappworth in England (1962) revealed informed consent was not always being followed in the United States and England. [70, 71] Researchers, who were not Nazis and considered themselves ethical, either ignored or were unaware of the Nuremberg Code. These revelations, together with the Tuskegee syphilis study scandal described in the next chapter, led to the creation of a commission and a federal system providing legal standards for evaluating the ethics of research experiments.

# 3.5: The Tuskegee Study & Belmont Report

FROM 1932 TO 1972 THE U.S. PUBLIC HEALTH SERVICE (USPHS) studied the natural course of untreated syphilis by following 399 African American men in Alabama who had the disease. In what became known as the Tuskegee Study, USPHS chose a county with a high syphilis rate and did not treat study participants with penicillin even after it was discovered and proven effective in 1947. According to the director of the Venereal Diseases unit of the USPHS from 1943 to 1948:

> "The men's status did not warrant ethical debate. They were subjects, not patients; clinical material, not sick people."[72]

By 1972, the Civil Rights movement had challenged racism and the second-class citizenship accorded to African Americans. There was widespread outrage across the country when the Tuskegee Study made headlines.

After the story broke, Congress held hearings and appointed a National Commission for the Protection of Human Subjects. The Commission published a series of important studies culminating in The Belmont Report in 1979. The Belmont Report laid down the principles and practices which have governed Research Ethics in the United States ever since.

Rather than the prevailing four principles approach to clinical ethics, The Belmont Report specified three principles for research ethics: Respect for Persons, Beneficence, and Justice.

*Respect for Persons* is similar to Autonomy in the four principles of clinical ethics. It emphasizes the importance of voluntary consent for a person to be in a study as a research participant. *Beneficence* in research ethics has the same meaning as in clinical ethics, that is promoting the well being of others. "Others," in this case, are future patients, not current patients or research participants. In the Belmont Report, Non-maleficence is not a separate principle but folded into the principles of Beneficence (including risks and benefits in any assessment) and *Justice* (protecting vulnerable populations from exploitation). The principle of Justice in research ethics is concerned with the question of who benefits and who bears the burdens of the research. In clinical ethics, Justice focuses on allocation of and access to medical resources. In research ethics Justice is primarily concerned with protecting underserved and vulnerable populations.

The Belmont Report suggests we implement each of the principles in the following ways:

Respect for Persons requires obtaining informed consent of research participants before enrolling them in a study. Informed consent requires at least three things: (1) giving each person all the information they need to make an informed decision; (2) making sure the participant comprehends the information to weigh risks and benefits; and (3) knows the decision of enrolling is entirely voluntary. Consent cannot be encouraged by using euphemisms or leaving out important information. Omission is as unacceptable as deliberate deception.

Beneficence requires researchers make complete assessments of study risks and benefits to those who enroll. A researcher must estimate the probability and magnitude of each harm and benefit the research could cause, and consider less risky alternatives to test the same hypothesis. All the risks must be minimized, remaining risks must be justified by the potential benefits from the research, and their likelihood must be disclosed in the informed consent process. If the research has unexpected outcomes, (i.e. adverse events such as unanticipated morbidity or mortality), then the IRB or a data safety monitoring board (DSMB) can stop the study at any time.

Justice must be observed by having a fair process to select participants, and fair outcomes. In describing the study, it should be clear what population could benefit from the research. Participants should be selected from that group if possible. There should not be unfair burden on vulnerable or disadvantaged individuals or groups simply because they are easier to find or recruit. The process used to identify and enroll participants should be equitable and should not exclude groups without good cause.

# 3.6: IRB: Consent & the Right to Withdraw

AFTER TUSKEGEE AND THE BELMONT REPORT, American society acknowledged biomedical research posed inherent ethical problems that needed to be addressed by setting national standards. Leaving decisions to the good intentions of individual scientists was no longer considered a reliable policy (as seemed implicit in the Nuremberg Code). Instead, the federal government created Institutional Review Boards (IRB) as the mechanism for ensuring the three principles outlined in the Belmont Report.

An **IRB** is a research ethics committee that reviews and approves any research proposal using human subjects before it can be carried out or submitted to a funding agency. The procedures and processes of IRBs are described in regulations known as 45 CFR 46, or "The Common Rule." CFR stands for Code of Federal Regulations, and the Common Rule refers to the fact that 19 different federal agencies use the same set of regulations.

IRBs require that a participant's consent must be documented on what is known as a Consent Form. It is common to have a preprinted form with all information and to ask participants to sign two copies. One copy is for the participant to take home, read, and formulate questions. The other is filed in the study's records. Participants must be told, and it is usually printed at the bottom of the consent form, that they may withdraw from the study at any time without fear of reprisal.

The consent form must include full disclosure of any reasonably foreseeable risks — physical, psychological, or social. The protocol should describe what steps will be taken in the informed consent process to ensure participants' understanding. For example, forms will usually be written for an eighth-grade reading level and printed in Spanish and English (or whatever the study population's prevalent languages are). Participants must understand that a study involves research, not treatment, and that if the PI is a physician, they are acting primarily as a scientist or researcher rather than as a medical doctor. Consent of participants must be voluntary, not the result of deception or coercion.

IRBs also identify potential conflicts of interest, that is, reasons why the PI may be biased towards wanting one result rather than another. They require that all COIs be disclosed and may require a PI respond in a way that manages potential COIs. An IRB's discretion can range from not approving research to having a PI disclose their financial interests (e.g. stock or employment by the company that manufactures a drug being tested) on the consent form. The greatest danger is a financial arrangement in which compensation to the PI for conducting the study could be influenced by study results (i.e., they get more money for some results than for others).

While IRBs have done a good job reducing some research injustices common until the 1970s, they are still controversial. Some believe IRBs are too bureaucratic and justify their existence by setting up unnecessary roadblocks to well-meaning researchers. Some believe studies that pose no risk shouldn't need IRB review at all. This claim is common among social science researchers, who invented the term "IRB creep" or "ethics creep" to describe it. [73]

Public interest in clinical research became more sympathetic during the 1980s and 90s, which some consider a possible paradigm shift. In the 1992 movie *Lorenzo's Oil*, the parents of a child with a rare life-threatening disease try to find a cure as his condition worsens. It's a true story told from the perspective of the parents. They were frustrated with the slow pace of research and what felt like roadblocks set up by research ethics. ACT-UP began in 1987, a group of gay rights activists who felt the U.S. was not spending enough money on research for a deadly new disease that was killing gay men (AIDS). [74] It was widely felt within the gay community that social bias led to a lack of sympathy with their plight and lack of interest in research to cure a new disease. Again, there was demand for more research, and questions about whether protecting participants was slowing progress.

Others feel, however, that even now IRBs are not powerful enough when it comes to biomedical research. For example, there is no prohibition against keeping studies "blind" (keeping study participants in the dark) after trends emerge. If early results indicate an intervention is either successful (a breakthrough) or dangerous, shouldn't the study be concluded and results publicized? Increasingly, IRBs are setting up independent DSMBs to oversee this. Good statisticians are especially important to make sure committees do not jump to conclusions if a sample is not statistically significant.

Some researchers are concerned there is no independent oversight body to review IRB decisions. The level of the training of members varies widely. A researcher whose protocol is rejected because of a misunderstanding may have no form of appeal. On the other hand, if a study is rejected by one IRB, there is no prohibition from taking it to another IRB. Since some IRBs are private for-profit entities, they may have a conflict of interest of their own and become known as "easier" IRBs from which to get approvals.

Lastly, there is no required compensation plan for injury in clinical trials, which leaves research participants financially vulnerable. Some believe a national no-fault compensation system should be instituted.[75]

# 3.7: Therapeutic Misconception

WHEN A PERSON HAS A SERIOUS ILLNESS and is enrolled in a clinical trial to test a new drug for that illness, it is almost inevitable they will harbor secret hopes of a "miracle," of being cured. It would be easy to take advantage of that person even without intending to do so. This is why IRBs spend a lot of time reviewing the wording of Consent Forms.

Common issues discussed by the members of an IRB as they review a proposed research protocol are: Is it misleading to call the PI "the doctor" and a research participant "a patient"? Should the drug being tested be called "medicine," "treatment," or "therapy"? The terms might mislead a patient into a false sense of security, an unspoken belief that the PI will put the patient's need for treatment ahead of the search for confirmation or disconfirmation of the hypothesis being tested. These feelings and beliefs are known as "therapeutic misconception," which can be seen as abusing the trust people have in doctors.

A researcher has a primary obligation to help future patients by discovering new knowledge, not helping the participants enrolled in their study. This is true whether the PI is an MD, DDS, or PhD. It is important to make the purpose of the study clear to potential participants. Attention to language is also why some names of some large multi-center trials are considered misleading, such as ALIVE, BEST, MIRACL, BRAVO, AFFIRM.[76]

Physicians and dentists must be sensitive to the dangers of playing two roles, clinician and researcher. They must find ways to deter the conflict of interest from leading to undue influence or subtle coercion. It must be made very clear to participants who are patients that they will not lose their doctor, nurse practitioner, or dentist if they decline to participate in the study. They must also know that they can withdraw from the study at any time without fear of losing their doctor or dentist. Some research centers require these messages be given to the patient by someone other than the PI if the PI happens to also be the patient's care provider. These are sometimes called "consent monitors" and may be approved by an IRB to ensure no bias towards enrolling persons into a trial.

# 3.8: Placebo-controlled Trials

MANY PEOPLE HAVE AN INTUITIVE DISTRUST of placebo-controlled studies. A placebo-controlled study is a way of testing a treatment by comparing those who receive it to a control group which receives a *placebo*, something intended to have no physiological effect. A proportion of the people who enroll in such studies get no effective treatment for their condition, which is considered unethical by some. Regulatory agencies in countries other than the U.S. tend to put less emphasis on placebo-controlled studies, as does the Helsinki Accord, which governs research ethics in most of the world, much like the Belmont Report does in the U.S.

For drug testing protocols in the U.S., however, a randomized, double-blind, placebo-control trial is often referred to as the gold standard. In a double blind study, neither participants nor researchers know which group is getting the treatment being tested and which group is receiving a placebo. Participants are randomly assigned to each group, hence the term "randomized control trial" (RCT). Results of randomized, double-blind, placebo-controlled studies are considered the best evidence for causal efficacy, ruling out the possibility that a small improvement is only the result of "the placebo effect." Placebo trials also have the advantage of giving statistically valid results with the fewest number of research participants, thus allowing researchers to get answers to questions more quickly than in other protocol methodologies. In situations where there is an urgent need for results or few potential study participants, placebo controls have an argument in their defense.

Placebo controls can be ethically justified if scientists aren't sure which of the two arms of their study is more effective. This is called "clinical equipoise," an important concept in research ethics.

Many large epidemiological studies are done abroad. Most European countries do not require placebo-controlled studies and have a national health service with national databases, making large epidemiological studies easier than in the U.S. The suggestion of a U.S. national database has lead to suspicion of government access to individual medical information.[77]

# 3.9: Vulnerable Populations

VULNERABLE POPULATIONS REQUIRE additional protection from potential research exploitation. These include those who either lack capacity to make decisions for themselves (e.g. young children, or patients with dementia) or lack the right to do so (e.g. prisoners and military personnel). Vulnerable populations receive special protections in government regulations. Before recent regulations, much research was done on minority groups, prisoners, soldiers, and the mentally ill or disabled.

HHS regulations 45 CFR 46 includes four subparts: A (the "Common Rule") governs IRBs; B, which provides additional protections for pregnant women, human fetuses, and neonates; C, which provides additional protections for prisoners; and D, which provides additional protections for children. Almost all federal agencies follow the Common Rule and usually adhere to other subparts if relevant to the agencies' areas of responsibilities. The typical requirement is that if a protocol poses more than a minimal risk (usually defined as what one normally encounters every day), then experts concerned with protecting the interests of that population should be included in the approval process.

# 3.10: International Research

MUCH HAS CHANGED in research ethics involving human participants since the advent of IRBs. However, whenever one issue is addressed in a way that makes it seem settled, another issue arises. Ethics is constantly challenged, always in the process of development, and never finished. In that way, it is much like science.

One of the more recent controversial issues involves international research. If a researcher conducts a study in another country, should it require IRB approval, and if so, should the IRB use the same standards as if the research was to be done in the U.S.? Many would argue the answer should be "yes" to both questions, perhaps based on an intuitive understanding of universal human rights. Not all would agree, though. Some critics would suggest every country has its own standards, customs, and ethics. For those critics, what we call "universal human rights" may seem more like Western presumption, ethical imperialism. Many believe Western nations have no right to impose their ethics on other cultures and societies.

Research that brought the issue to light concerned AIDS clinical trial group number 76 (ACTG-076). It was known in the U.S. that standard azidothymidine (AZT) therapy reduces maternal-child HIV transmission by 66% (from 25% to 8%).[78] AZT is very expensive, so a trial was done in Africa to test using lower doses of AZT ($80 worth instead of $800). The control group received a placebo, justified because cost of standard treatment was so high it was not available in Africa.

This study would never be approved by an IRB if it were done in the U.S. Nevertheless, results showed the lower dose reduced transmission by 50% — a significant finding.[75]

ACTG-076 raised many questions. Should therapy offered to a control group be a local standard or an international standard? Was the study taking advantage of lower standards in Africa to save money and to speed up results? Was this study really Tuskegee II, displacing racism from African Americans to Africans? Other cultural issues raised by the African AIDS trial are related to the anthropological notion of cultural relativism. Is informed consent itself a Western presumption that does not apply universally? Why should we expect written consent for research in countries with no tradition of consent in medicine, or from women, if they don't make other medical decisions in that country? Should we follow local custom and allow consent by a woman's husband or village elder?

Although these questions are still being debated, the current consensus seems to be that as a sign of respect for some cultures, it is acceptable to seek a community leader's consent before talking to individuals and seeking their consent. This may require hiring local people to translate and explain terms like "disease," "causation," "treatment," "placebo," and "study."

Judging research in developing nations by U.S. ethical standards may appear unfair. Our acknowledgment of past errors led to progress in ethics. While we must be fair in our expectations of other places, it is better to raise ethical and medical standards where research is being done than to take advantage of lower local standards.

International patent law and trade agreements can include exceptions that allow countries to legally void a patent for their population if there's a public health emergency. This often leads to a compromise where a drug company lowers its price in that country. It is unethical for a population to be used to get a drug approved that they cannot afford to buy.

The Council for International Organizations of Medical Sciences (CIOMS) proposes a host country should always benefit from research done in their country. This might include sharing profits with a company doing research, providing drugs for free or at reduced cost, or by getting other health benefits such as building a research clinic that can later be used by the country's public health department.

CIOMS also asks researchers to help countries strengthen ethics review infrastructures and research capacity where they do perform clinical trials. For international students, ethics courses like this can be a valuable subject you can teach in your home country.

# 3.11: Humane Use of Animals in Research

UNTIL RECENTLY, concern for animal welfare was the province of a few. Since the mid-to-late twentieth century, animal welfare has become an increasingly important issue among researchers and the public. Perhaps awareness that animals have consciousness has undermined the traditional lack of concern, based on the view espoused by Rene Descartes in the late 1600s that people have a soul and animals do not.[79] Today, there is growing concern for the welfare of animals used in research.

With his book *The Origin of Species* (1859), Charles Darwin laid out scientific evidence for his theory of evolution, which established humankind's place in nature. At about the same time, philosopher Jeremy Bentham initiated ethical inquiry into human use of animals, when he asked, "The question is not can they think? But can they suffer?" *Speciesism* is a recently coined word to describe treating some species differently from others. The term implies that this is unfair (analogous to racism or sexism) based on a view of the moral standing of animals. However, for scientific researchers exploring clinical problems, the closer an animal is to humans in structure and function, the more likely it is to provide relevant and reliable clinical information. Those opposed to speciesism would say the closer an animal is to humans the greater the moral concern it deserves and the greater our obligation to not misuse it.

Research on primates, the closest living relatives to humans, is especially controversial. In the 1990s, secret videotapes were made of research involving head trauma inflicted on restrained chimpanzees at the University of Pennsylvania.[80] Recordings of researchers laughing at frightened chimps reinforced the fear that animal researchers do not care about animals' pain or suffering. Belief that animals genetically close to humans deserve special protection has led some countries (including much of Europe, Australia, and Japan) to outlaw all research on Great Apes, for example.

Some of the worst publicity about animal research involves reported use of dogs bought from pounds or people who collect strays. There have been stories of people whose pet dog got lost and was picked up by the local dog catcher.[81] The pound then sold the dog to a laboratory to be used in research and then euthanized. This practice can make researchers look unethical, all the more so because it has purportedly motivated some people to steal dogs to sell for research. Some animal welfare advocates have called for a ban on the use of any dogs or cats that come from "Class B" dealers. Many people in the U.S. view pounds as shelters or havens for homeless animals, not a resource for biomedical research. The practice of using animals from pounds potentially endangers public support for research.[82]

It is important to keep in mind that about 95% of animals used in research are mice and rats that were bred for just that purpose.[83] All animals should be well treated in our care and sacrificed in ways that minimize pain and suffering. If anything, this should be even more important for higher mammals.

Objective science doesn't mean value-free science. Ethical concerns cannot be disregarded. Even if we derive important benefits from animal research, there is still a question of when animal research is ethically justifiable. Scientific progress is an important value, but not the only important value. When reviewing a proposed experiment, the first question is whether the goal of the proposed study is worthwhile. The second is whether proposed methods will answer the hypothesis. Ethics adds another question, one not answered by science alone: Are the methods ethical? In research ethics, knowledge and values are both essential and neither can replace the other. We require ethical methods to be an integral part of scientific progress.

# 3.12: Institutional Animal Care & Use Committee

IN BOTH HUMAN AND ANIMAL RESEARCH, a committee with the authority to require changes or reject a study is now required to review research protocols before funding.

For animal research, ethics review committees are called an Institutional Animal Care and Use Committee (IACUC). All researchers and lab workers using animals must be trained on the humane handling of animals. IACUCs have veterinarians to supervise care of the animals, a field of veterinary medicine known as laboratory animal medicine. IACUCs review all proposed animal research. Their three principles, analogous to the three principles of the Belmont Report, are called "the three Rs": Reduce, Replace, and Refine.

*Reduce* means researchers should use only as many animals as are needed to get valid results, and no more. Statistical requirements mean researchers should use as many animals as needed to get valid results, and no less. IACUCs should have a biostatistician on the committee to help with this task.

*Replace* means animals should not be used if results on the same issue can be obtained without use of animals. Advances in cell biology, genomics, proteomics, and other fields mean much basic science research is done at a cellular or genetic level and does not require live animals.

*Refine* means shaping a protocol to minimize pain and suffering for animals. Researchers should use the species that has the least sensitivity to pain and suffering or has the fewest needs to lead an acceptable life. If research can be done using an invertebrate, then it is not justified to use a vertebrate. If a vertebrate is necessary, reptiles and amphibians should be used unless the research requires mammals. Small mammals are to be used instead of higher mammals when possible. Adequate anesthesia must always be used. To minimize suffering, everyone who handles research animals must have adequate training to work and build rapport with them (especially mammals and birds).

If a student expresses discomfort about animal research, this may indicate they are sensitive to the ethical issue. Such a student might be encouraged to work with *Drosophila*, *C. elegans*, *Aplysia*, and yeast. Scientists with an interest in ethics might volunteer on an IACUC to help ensure regulations are followed. This demonstrates that an IACUC is as important ethically as an IRB. Even for those to whom these issues do not seem as important as human research, it is still important to understand that in research ethics and regulatory requirements, animals and IACUCs are analogous to human participants and IRBs.

# ABBREVIATIONS

| | |
|---|---|
| AAA | American Automobile Association |
| AAAS | American Association for Advancement of Science |
| AAMC | Association of American Medical Colleges |
| ABMS | American Board of Medical Specialties |
| ACE | American College of Epidemiology |
| ACOG | American Congress of Obstetricians and Gynecologists |
| ACP | American College of Physicians |
| ADA | American Dental Association |
| AFFIRM | Atrial Fibrillation Follow-up Investigation of Rhythm Management study |
| AHRQ | Agency for Healthcare Research and Quality |
| AIDS | Acquired Immune Deficiency Syndrome |
| ALD | Adrenoleukodystrophy |

| | |
|---|---|
| ALIVE | AIDS Link to Intravenous Experience study |
| AMA | American Medical Association |
| AMIA | American Medical Informatics Association |
| AMSA | American Medical Student Association |
| ANA | American Nurses Association |
| AAP | American Academy of Pediatrics |
| ASM | American Society for Microbiology |
| AZT | Azidothymidine |
| BEST | Beta-Blocker Evaluation of Survival Trial |
| BRAVO | Breast Reduction Assessment: Value and Outcomes study |
| CAM | Complementary and Alternative Medicine |
| CAT Scan | Computerized Axial Tomography, also called CT scan |
| CDC | Centers for Disease Control and Prevention |
| CFR | Code of Federal Regulations |
| CIOMS | Council for International Organizations of Medical Sciences |

| | |
|---|---|
| CME | Continuing Medical Education |
| COI | Conflict of Interest |
| CQI | Continuous Quality Improvement |
| CTSA | Clinical and Translational Science Awards |
| CVMBS | Colorado State University College of Veterinary Medicine & Biological Sciences |
| DDS | The academic degree of Doctor of Dental Surgery |
| DHHS | Department of Health and Human Services |
| DMC | Decision-making Capacity |
| DNP | The academic degree of Doctor of Nursing Practice |
| DSMB | Data Safety Monitoring Board |
| DTC | Direct to Consumer |
| DWI | Driving While Intoxicated |
| EAP | Employee Assistance Program |
| EHR | Electronic Health Records |
| EMR | Electronic Medical Records |
| ENT | Ear, Nose & Throat |
| FDA | Food and Drug Administration |

| | |
|---|---|
| FFP | Fabrication, Falsification, and Plagiarism |
| GI | Gastrointestinal |
| GPS | Global Positioning System |
| HHS | Department of Health & Human Services |
| HIPAA | Health Insurance Portability and Accountability Act |
| HIV | Human Immunodeficiency Virus |
| IACUC | Institutional Animal Care and Use Committee |
| ICU | Intensive Care Unit |
| ID | Identity |
| IOM | Institute of Medicine |
| IRB | Institutional Review Board |
| JAMA | Journal of the American Medical Association |
| MBA | The academic degree of Masters of Business Administration |
| MD | The academic degree of Medical Doctor |
| MIRACL | Myocardial Ischemia Reduction with Aggressive Cholesterol Lowering study |

| | |
|---|---|
| MPH | Master of Public Health |
| MS | Medical student |
| NAACP | National Association for the Advancement of Colored People |
| NAS | National Academy of Sciences |
| NBME | National Board of Medical Examiners |
| NIH | National Institutes of Health |
| NSF | National Science Foundation |
| ORI | Office of Research Integrity of the Department of Health and Human Services |
| OSHA | Occupational Safety and Health Administration |
| PhD | The academic degree of Doctor of Philosophy |
| RCR | Responsible Conduct of Research |
| RCT | Randomized Controlled Trial |
| QA | Quality Assurance |
| PI | Principal Investigator |
| RSVP | From the French, *répondez s'il vous plaît*, request for response. |
| SCCM | Society of Critical Care Medicine |

| | |
|---|---|
| sCHIP | State Children's Health Insurance Program |
| STD | Sexually Transmitted Disease |
| STI | Sexually Transmitted Infection |
| TBL | Team-based Learning |
| Z-Pak | Package of Zithromax antibiotic tablets |

# REFERENCES

[1] Preston J. Social media history becomes a new job hurdle. July 20, 2011. New York Times. URL: http://www.ada.org/sections/advocacy/pdfs/ada_workforce_statement.pdf. Accessed July 21, 2011.

[2] www.Plagiarism.org

[3] May T. A guide to avoiding plagiarism (2nd ed.), Austin, Texas: LBJ School of Public Affairs, The University of Texas Austin.

[4] American Nurses Association. Code of Ethics for Nurses. URL: http://www.nursingworld.org/MainMenuCategories/EthicsStandards/CodeofEthicsforNurses.aspx. Accessed July 19, 2011.

[5] American Dental Association: ADA Principles of Ethics and Code of Professional Conduct. URL: http://www.ada.org/194.aspx. Accessed July 19, 2011.

[6] American Medical Association: AMA's Code of Medical Ethics. URL: http://www.ama-assn.org/ama/pub/physician-resources/medical-ethics/code-medical-ethics.page. Accessed: July 19, 2011.

[7] Barfett J, Lanting B. Pharmaceutical marketing to medical students: the student perspective. MjM 2004 8:21-17.

[8] Sufrin CB, Ross JS. Pharmaceutical industry marketing: understanding its impact on women's health. Obstet Gynecol Surv 63 (9): 585–96.

[9] Aggarwal K. The relationship between pharmaceutical companies and physicians. Berkeley Scientific Journal, 2010;13(2).

[10] Moghimi Y. The "PharmFree" campaign: educating medical students about industry influence. PLoS Med. 2006 Jan;3(1):e30. Epub 2006 Jan 31.

[11] Sierles FS, Brodkey AC, Cleary LM, McCurdy FA, Mintz M, et al. Medical students' exposure to and attitudes about drug company interactions: A national survey. JAMA. 2005;294:1034–1042.

[12] American College of Obstetricians and Gynecologists Committee on Ethics, Opinion No. 385, Nov. 2007.

[13] Schloendorff v. The Society of the New York Hospital (105 N.E. 92) 1914.

[14] Salgo v. Leland Stanford, etc. Bd. Trustees 154 Cal.App.2d 560.

[15] U.S. National Institutes of Health, Office of Human Subjects Research. Nuremberg Code. Accessed on July 17, 2011.

[16] Cohen LM, Germain MJ, Poppel DM, Woods AL, Pekow PS, Kjellstrand CM. Dying well after discontinuing the life-support treatment of dialysis. Arch Intern Med. 2000 Sep 11;160(16):2513-8.

[17] Prince v. Massachusetts, 431 U.S. 494 (1977).

[18] Committee on Bioethics, American Academy of Pediatrics. Guidelines on Forgoing Life-Sustaining Medical Treatment. Pediatrics 93;3 (March, 1994): 532-536.

[19] Resolution on "Conscientious Objection." International Federation of Gynecology and Obstetrics. URL: http://www.figo.org/projects/conscientious. Accessed: July 19, 2011.

[20] Novack DH, Plumer R, Smith RL, Ochitill H, Morrow GR, Bennett JM. Changes in physicians' attitudes toward telling the cancer patient. JAMA. 1979 Mar 2;241(9):897-900.

[21] President's Commission for the Study of Ethical Problems in Medicine and in Biomedical and Behavioral Research, Making Health Care Decisions, Washington: U.S. Government Printing Office, 1982.

[22] Kohn LT, Corrigan JM, Donaldson M. ed.s. To err is human: building a safer health System. Washington, D.C: National Academy Press, 2000.

[23] Gallagher TH, Waterman AD, Ebers AG, et al. Patients' and physicians' perspective regarding the disclosure of medical errors. Journal of the American Medical Association 2004; 289: 1001-1007.

[24] Beckman HB, Markakis KM, Suchman AL, Frankel RM. The doctor-patient relationship and malpractice. Lessons from plaintiff depositions. Arch Intern Med. 1994 Jun 27;154(12):1365-70.

[25] Council on Ethical and Judicial Affairs, AMA, Conflicts of Interest: physician ownership of medical facilities, JAMA 1992; 267: 2366-2369.

[26] Mitchell JM. Prevalence of physician self-referral after Stark II. Health Affairs 2007: 415-424.

[27] Kouri BE, Parsons RG, Altert HR. Physicians self-referral for diagnostic imaging: Review of the Empiric Literature. American Journal of Radiology 2002;179:843-850.

[28] Wazana A. Physicians and the pharmaceutical industry: is a gift ever just a gift? JAMA 2000; 283:373-380.

[29] Steinbrook R. Commercial support and CME. NEJM 2005; 352: 534-535.

[30] Dana J, Loewenstein G. A social science perspective on gifts to physicians from industry, JAMA, 2003:290: 252-255.

[31] Frosch DL, Grande D. Direct-to-consumer advertising of prescription drugs. LDI Issue Brief. 2010 Mar-Apr;15(3):1-4.

[32] Lind SE. Finder's fees for research subjects. NEJM 323 (1990): 192-195.

[33] Chimonas S, Kassirer JP. No more free drug samples? PLoS Med 6(5): e1000074. doi:10.1371/journal.pmed.1000074.

[34] Morelli DM, Koenignsberg MR. Sample medication dispensing in a residency practice. 1992 J Fam Pract 34: 42–48.

[35] Westfall JM, McCabe J, Nicholas RA. Personal use of drug samples by physicians and office staff. 1997 JAMA 278: 141–143.

[36] World Health Statistics 2009. World Health Organization. May 2009. Accessed July 20, 2011.

37 When in trouble, doctors flee to medicine-land. BMJ. 2005 January 22; 330(7484): 0.

38 Bodenheimer T.S., Grumbach K. Understanding Health Policy: A Clinical Approach, 2nd ed. 1998. New York: McGraw-Hill.

39 Mullen J. European Association for Paediatric Dentistry. History of water fluoridation. Br Dent J. 2005 Oct 8;199(7 Suppl):1-4.

40 Deer B. Revealed: MMR research scandal. The Sunday Times. February 22, 2004. London. URL: http://www.timesonline.co.uk/tol/news/uk/health/article1027603.ece. Accessed July 18, 2011.

41 Triggle N. MMR scare doctor "acted unethically," panel finds. January 28, 2010. BBC News. URL: http://news.bbc.co.uk/2/hi/health/8483865.stm. Accessed July 18, 2011.

42 Meikle J, Boseley S. MMR row doctor Andrew Wakefield struck off register. May 24, 2010. The Guardian (London). URL: http://www.guardian.co.uk/society/2010/may/24/mmr-doctor-andrew-wakefield-struck-off. Accessed July 18, 2011.

43 Institute of Medicine. For the public's health: the role of measurement in action and accountability. 2011. Washington, DC: The National Academies Press.

44 Ropeik D. Public health: Not vaccinated? Not acceptable. July 18, 2011. Los Angles Times. URL: http://www.latimes.com/news/opinion/commentary/la-oe-ropeik-vaccines-20110718,0,4240440.story. Accessed July 19, 2011.

45 Abelson R. Census numbers show 50.7 million uninsured. New York Times. URL: http://prescriptions.blogs.nytimes.com/2010/09/16/census-numbers-show-50-million-uninsured/. Accessed July 20, 2011.

46 McCann K. Medicaid Reform. Texas Academy of Family Physicians. URL: http://www.tafp.org/news/tfp/07no4/legeUpdate.asp. Accessed July 20, 2011.

[47] Selden TM, Sing M. The distribution of public spending for health care in the United States, 2002. Health Aff (Millwood). 2008 Sep-Oct; 27(5):w349-59. Epub 2008 Jul 29.

[48] Connelly J. Doctors are opting out of Medicare. April 1, 2009. New York Times. URL: http://www.nytimes.com/2009/04/02/business/retirementspecial/02health.html. Accessed July 19, 2011.

[49] Ethical Issues in Health Care Reform. JAMA. 1994 Oct 5;272(13): 1056-62.

[50] Institute of Medicine. The future of nursing: leading change, advancing health. 2011 Washington, DC: The National Academies Press.

[51] Bertakis KD, Azari R. Patient-centered care is associated with decreased health care utilization. J Am Board Fam Med. 2011 May-Jun;24(3):229-39.

[52] The State of telemedicine and telehealth in Texas. Texas Statewide Health Coordinating Council. URL:www.dshs.state.tx.us/chs/shcc/reports/tmreport.pdf. Accessed July 20, 2011.

[53] Cost of delay. Pew Center of the States. 2010. URL: http://www.pewcenteronthestates.org/uploadedFiles/Cost_of_Delay_web.pdf. Accessed July 20, 2011.

[54] Breaking down barriers to oral health for all Americans: The role of workforce. American Dental Association. 2011. URL: http://www.ada.org/sections/advocacy/pdfs/ada_workforce_statement.pdf. Accessed July 21, 2011.

[55] Improving Access to Oral Health Care for Vulnerable and Underserved Populations. http://www.iom.edu/Reports/2011/Improving-Access-to-Oral-Health-Care-for-Vulnerable-and-Underserved-Populations.aspx. Accessed January 4, 2012.

[56] Fries JF, Koop CE, Beadle CE, Cooper PP, England MJ, Greaves RF, Sokolov JJ, Wright D. Reducing health care costs by reducing the need and demand for medical services. The Health Project Consortium. N Engl J Med. 1993 Jul 29;329(5):321-5.

[57] Ezzati M, Friedman AB, Kulkarni SC, Murray CJ. The reversal of fortunes: trends in county mortality and cross-county mortality disparities in the United States. PLoS Med. 2008 Apr 22;5(4):e66.

[58] Jemal A, Ward E, Anderson RN, Murray T, Thun MJ. Widening of socioeconomic inequalities in U.S. death rates, 1993-2001. PLoS One. 2008 May 14;3(5):e2181.

[59] http://www.who.int/social_determinants/en/

[60] http://www.cdc.gov/socialdeterminants/

[61] National Healthcare Disparities Report, 2003. Agency for Healthcare Research and Quality, Rockville, MD. http://www.ahrq.gov/qual/nhdr03/nhdr03.htm.

[62] International Committee of Medical Journal Editors. Uniform requirements for manuscripts submitted to biomedical journals: writing and editing for biomedical publications. http://www.icmje.org/. Accessed July 19, 2011.

[63] Shamoo A, Resnik D. Responsible conduct of research. 2nd ed. (p. 142). New York, NY: Oxford University Press; 2009.

[64] Maugh TH, Mestel R. Key breast cancer study was a fraud. April 27, 2001. Los Angles Times. URL: http://articles.latimes.com/2001/apr/27/news/mn-56336. Accessed July 11, 2011.

[65] Cauvin HE. Cancer researcher in South Africa who falsified data is fired. New York Times (Print). 2000 Mar 11:A16.

[66] Shamoo A, Resnik D. Responsible conduct of research. 1st ed. (p. 44). New York, NY: Oxford University Press; 2003.

[67] Office of Research Integrity. URL: http://ori.hhs.gov/TheLab/. Accessed July 19, 2011.

[68] Annas GJ, Grodin MA, editors. The Nazi doctors and the Nuremberg Code: human rights in human experimentation. New York: Oxford University Press.

[69] Trials of War Criminals before the Nuremberg Military Tribunals under Control Council Law No. 10, Vol. 2, pp. 181–182. Washington, D.C.: U.S. Government Printing Office, 1949.

[70] Beecher H. Ethics and clinical research. From the anaesthesia laboratory of the Harvard Medical School at the Massachusetts General Hospital. 1966. Int Anesthesiol Clin. 2007 Fall;45(4):65-78.

[71] Pappworth MH. Human guinea pigs: a warning. (1962) Twentieth Century Magazine: 66–75.

[72] Jones, J. Bad blood: The Tuskegee syphilis experiment: A tragedy of race and medicine. (1981) NY: The Free Press.

[73] Haggerty KD. Governing social science research in the name of ethics. Qualitative Sociology, Vol 27, No. 4, Winter 2004.

[74] ACT-UP. URL: http://www.actupny.org/. Accessed July 19. 2011.

[75] Steinbrook R. Compensation for injured research subjects. N Engl J Med. 2006 May 4;354(18):1871-3.

[76] Hochhauser M. Therapeutic Misconception and Recruiting Doublespeak in the Informed Consent Process pp. 221-223, in Ezekiel Emanuel et al., eds., Ethical and Regulatory Aspects of Clinical Research: Readings and Commentary. 2003. Baltimore: The Johns Hopkins University Press.

[77] Delaney A. Hopes and fears in DNA databases. ABC News. June 30, 2006. URL: http://abcnews.go.com/US/LegalCenter/story?id=2139524&page=1. Accessed July 19, 2011.

[78] Shamoo A, Resnik D. Responsible conduct of research. 1st ed. (p. 330). New York, NY: Oxford University Press; 2003.

[79] Anscombe E, Geach PT. Descartes philosophical writing. Indianapolis, IN: Bobbs-Merrill; 1971.

[80] Shamoo A, Resnik D. Responsible conduct of research. 1st ed. (p. 232). New York, NY: Oxford University Press; 2003.

[81] Gillham C. Bought to be sold. Newsweek. February 17, 2006. URL: http://www.newsweek.com/2006/02/16/bought-to-be-sold.html. Accessed July 20, 2011.

[82] Shelter Derived Animal Use Guidance Statements. Colorado State University College of Veterinary Medicine and Biomedical Sciences URL: http://www.cvmbs.colostate.edu/cvmbs/ ShelterAnimalUse.htm. Accessed July 20, 2011.

[83] American Anti-Vivisection Society. Mice and rats used in research. URL: http://www.aavs.org/site/c.bkLTKfOSLhK6E/b.6456937/k. 1F85/Mice_and_Rats_Used_in_Research.htm. Accessed July 19, 2011.

CPSIA information can be obtained at www.ICGtesting.com
Printed in the USA
LVOW09s2233010914

401799LV00007B/1071/P